D0629346

A series of student texts in

CONTEMPORARY BIOLOGY

General Editors:
Professor E. J. W. Barrington, F.R.S.
Professor Arthur J. Willis

The Biology
of the Arthropoda

Kenneth U. Clarke
D.Sc.
Reader in Zoology, University of Nottingham

American Elsevier Publishing Company, Inc.
New York

American Elsevier Publishing Company, Inc.
52 Vanderbilt Avenue, New York, N.Y. 10017

First published in Great Britain by
Edward Arnold (Publishers) Ltd.

ISBN: 0-444-19559-9
Library of Congress Catalog Card Number: 73-2309

Printed in Great Britain by
William Clowes & Sons, Limited
London, Beccles and Colchester

Preface

The object of zoological science is to seek out the laws and principles which govern the expression of animal life on this earth and through their application to understand the possibilities, limitations and diversity of the organisms which constitute this planet's fauna. This book is an attempt to do this for one of the major groups of the Animal Kingdom, the Phylum Arthropoda.

This requires a very different approach from that found in most textbooks. The basic factual matter that has been selected is somewhat different and is drawn from a very wide field of biological studies. A good deal of theoretical material is included and it has been necessary to devise ways of expressing in an orderly and reasoned manner the many diverse properties of the individual animal and of higher groupings up to Phylum level. The applicability of these methods to a wider and more detailed array of facts than those given in this book has been explored and found satisfactory, so that more advanced studies can be founded upon them and they may be applied to other types of animal organization.

This book is intended for students in their first and second years at a University and gives them a basis of facts and thought about arthropods which can be expanded in scope and depth in future studies. It is not perhaps an easy book, but then zoology is not an easy science. The book however makes no greater demands on the student than do the texts of, say, physics, chemistry or mathematics, commonly used in their first and second year studies and requires no greater preparation than that given in pre-university biological courses.

I should like to thank all those authors who have given permission to use figures from their papers; individual acknowledgements are noted beneath each illustration. Especially I would like to thank my wife, who typed the manuscript and gave much valuable help in its preparation, for her work which has so greatly facilitated the preparation of this book.

Nottingham   K.U.C.
1972

Table of Contents

I

Introduction

Arthropoda is a name given to a group of animals which, one way or another, express a design common to them all and significantly different from that shown by all other animals. In terms of numbers of individuals and species the arthropod design has proved to be a very successful one and most of the world's fauna show it. What this design is, what its major variations are, how it is expressed in the individuals governed by it, what its possibilities and limitations are, and what are its fundamental properties are questions that this book sets out to answer.

The arthropod design is one of many such designs found in animals whose existence is noted in zoological classification by allotting them the rank of Phyla. The faunas in which these designs evolved have long since passed from the world, leaving relatively little trace of their existence in fossil remains but giving rise to a number of fairly distinct Phyla whose members have constituted all succeeding faunas. All Phyla are not equally complex in design. Some, such as the Coelenterata, have a simple basic plan which must have survived from the early stages of evolution. Others, such as the Arthropoda and Chordata, are much more complex and have a long evolutionary history behind them.

The separate features of the arthropod design accumulated stage by stage in the evolutionary steps leading up to its appearance in the species that first displayed it. There is evidence that the design may have evolved more than once, the animals displaying it being derived from ancestors in which its features accumulated in different sequences and under different conditions. If this is so then the design, once it appeared, so dominated its possessors that they may well be considered as a single type, the separate origins simply serving to stress the effectiveness of the plan as a pattern for survival.

The vast variety of form and function displayed by the members of this Phylum can, for each individual, be analysed into systems (muscular, respiratory etc.), organs (heart, gonads etc.), tissues (epithelial, connective etc.), and cells (neurons, haemocytes etc.), each of which has properties peculiar to its particular level of organization. By reversing this analysis and considering first the properties of cells, then of tissues, organs, systems and finally, whole animals, it is possible to describe and understand the structure and function of a wide range of arthropod types through a synthesis which can be greatly extended to include material not dealt with in the present volume.

The resulting picture is of an organization considered over a fairly brief time period when any one organism can display its functions, but not alter greatly its design. Arthropods however have the ability to alter their form considerably during post-embryonic development; if necessary, several times. This aspect of arthropod organization is dealt with by considering the life patterns they possess.

Evolution within the arthropods is a very complex story and only the briefest outline, practically devoid of evidence, can be given here. However it plays an important part in our understanding of the group, not only for the obvious reasons, but because it shows again and again similar solutions developing in different evolutionary lines to meet similar circumstances. It is this relatively limited repertoire leading to diversity produced by different combinations of recognizable entities that makes a study of arthropod organization simpler than it might otherwise be.

By this stage a great deal of diverse information will have been presented and some means must be found of relating it together in a concise form. This is developed in the last chapter as a system of matrices which allows concise statements of the properties of individual, species, or higher categories of arthropods to be made. Finally, there is consideration of the fundamental forces that have been coupled with the structure, function and dynamics of the arthropod to give the design its possibilities and its limitations.

This book has been written as an integrated whole, the chapters to be read in the sequence in which they are given. Nevertheless it is possible to omit the chapters on cells or tissues or organs, or to study these in the inverse order if that approach is preferred, with however some loss of content. It should be noted that a general knowledge of cells and/or of tissue structure is not a substitute for these chapters: at these levels the arthropod differs somewhat from other organisms. These details, small when considered in isolation, are collectively important, for it is these in total which, together with the higher level designs, make the arthropod not merely another animal, but a unique biological organization.

2

The Cells of Arthropods

INTRODUCTION

The lowest analytical level considered here is that of the cells, which consist of a number of different types. Some of these types are very widespread throughout the animal kingdom; a few seem to have evolved only within the arthropods (tracheolar cells), while some which are widespread elsewhere (mucus and ciliary cells) are rare within, or absent from, the group. Complete identity of an arthropod cell, for example a neuron, with one from another Phylum cannot occur since within the neuron there is information unique to the arthropod design. Its functions however impress upon it a common pattern and in this respect arthropod neurons differ only in detail from those of other Phyla. The account given will serve to remind us of this close resemblance and also that, at this and all levels, properties common to all animals must be considered together with the special arthropod features if an understanding of the properties of the organism is to be attained.

GENERAL PROPERTIES

In arthropods the basic structure and function of the cells are very similar to those occurring in other animal groups. The main biochemical pathways follow a common pattern; such modifications as occur require little consideration at an elementary level and do not in any case account for the organization of cells into arthropods.

Since cells arise from pre-existing cells by nearly equal division the daughter cells will receive information in all their constituent parts. However, that contained on the chromosomes within the nucleus plays a

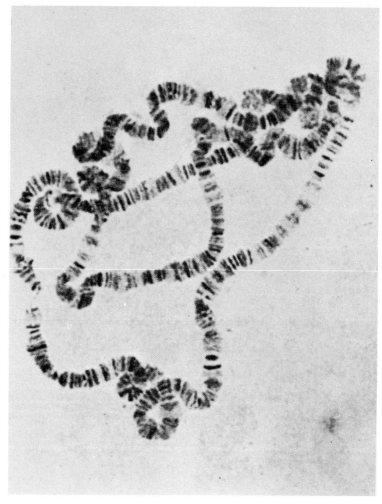

Fig. 2.1   The giant chromosomes from the salivary glands of *Drosophila*. (After Berendes, 1965, *Chromosoma*, **17**, 50)

dominant part in dictating the design and organization of an arthropod, and it is on this aspect that we will concentrate.

As in other animals, genetic information is coded in the structure of deoxyribose nucleic acid (DNA). Few determinations exist which permit a comparison of the amount of DNA per haploid nucleus in the different Phyla. Arthropods have a value intermediate between that found in other invertebrate Phyla and that for Chordata, presumably indicating that more information is necessary to produce an arthropod than other invertebrates, and/or that more of it is encoded upon the chromosomes.

Information contained upon the chromosomes is decoded as follows. Messenger ribonucleic acid (mRNA) is formed at the DNA molecule; it separates from the DNA and passes out of the nucleus into the cytoplasm where it becomes associated with the ribosomes. In the cytoplasm amino acid molecules become linked with transfer RNA (tRNA), a specific tRNA for each type of amino acid. The tRNA/amino complex passes to the ribosome where, under the influence of mRNA, the amino acids are assembled into a specific enzyme or protein. While exceptions occur, observation and experiment confirm that the entire genetic information content present in the zygote nucleus is replicated in each of the nuclei of all the cells of the body. In some Diptera the body cells contain very large chromosomes in which the genes are visible as numerous transverse bands along their length (Fig. 2.1). When one of these sites becomes active it swells up, returning to normal size when activity ceases. A comparison can be made between the patterns produced by these puffs of Balbiani within a nucleus at different times, or between nuclei of cells engaged in different activities. These patterns indicate that, for each type of cell and for each phase of its activity, there is a definite and constant arrangement of these puffs (Figs. 2.2 and 2.3). This implies that only a part of the entire genetic information is utilized within each cell. It is assumed that similar events occur within the cells of all other arthropods.

To some extent all cells that make up a multicellular organism are specialized, since not one of them can realize the full diversity of function encoded in the information they contain. Generalization and specialization are comparative terms. Generalized cells perform many functions each of which is known to be the main function of a specialized cell elsewhere. They include non-working cells not yet committed to a job in the cellular architecture of the animal. For example, the mid-gut cells of *Peripatus* are generalized cells since they produce digestive enzymes, absorb the products of digestion, synthesize them and store them until required. In higher arthropods cells occur which are specialized for each of these functions. Specialized cells are those such as nerve cells for the conduction of electrical impulses, rhabdome cells for the reception of light, or tracheolar cells producing the fine terminal tubules of the respiratory system. In addition

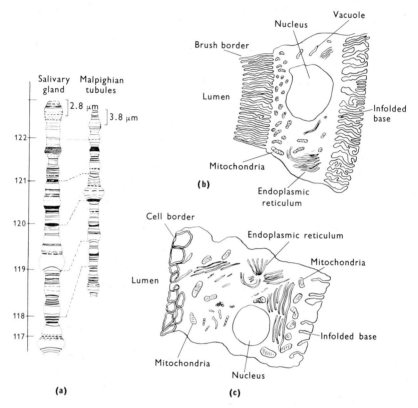

Fig. 2.2 Puffing patterns and cellular differentiation. (**a**) Differences in the puffing patterns of two short lengths of chromosome, one from the salivary glands and one from the Malpighian tubules of *Drosophila melanogaster* (Berendes, 1966). (**b**) Sketch of cells from the Malpighian tubules, and (**c**) from salivary glands showing general cytological differences.

to this dominant single function they also perform many of the normal maintenance and energy-producing reactions common to most cells.

Many molecules, atoms and sub-atomic particles pass through the cell membrane without disrupting it. Its permeability to these is similar to that of animal cells in general and is governed by the same controlling factors. Materials too large to pass through the intact membrane become enclosed in a small pocket formed from the cell membrane. The pocket detaches from the membrane which then establishes normal continuity behind it: finally the walls of the vesicle disappear, releasing its contents which are now on the opposite side of the cell membrane to that on which they were

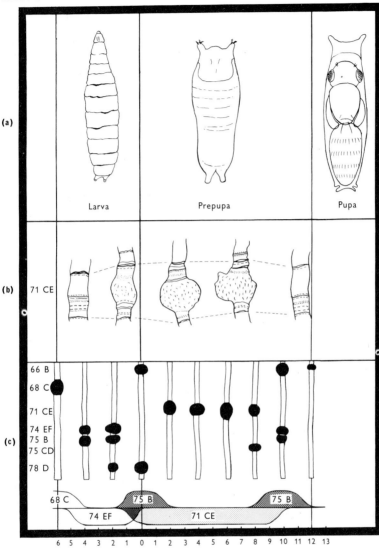

Fig. 2.3 Changes in puffing patterns during the development of *Drosophila melanogaster*. (**a**) Changes in external form, late larva, prepupa and early pupa. (**b**) Changes in the puffing of one locus 71 CE during this period as it appears in the actual specimens. (**c**) Diagram illustrating changes in a short length of one of the chromosomes: the figures identify the loci, time is given in days, zero being the moment of change from the larva to the prepupa; the graph indicates the duration of some of the puffs. ((**b**) and (**c**) redrawn from Becker, 1959)

when first engulfed by the vesicle. Material may enter the cell by this process (endocytosis) or leave it (exocytosis).

CELL TYPES

The major cell types of arthropods can be described under the following classes: the epidermal cell and its derivatives; the muscle cells; the secretory, absorptive, metabolic and excretory cells; the nerve cells and the generative cells.

The epidermal cell and its derivatives

These are the cells which secrete the cuticle, both on the body surface and lining the ducts which are ectodermal ingrowths; they produce the surface secretions and form the primary sense cells of the animal.

In its resting state the epidermal cell is thin and flattened and its boundaries with neighbouring cells difficult to distinguish. Its outer surface is closely applied to the cuticle and is produced into numerous fine protoplasmic filaments which penetrate the cuticle ending just beneath its thin

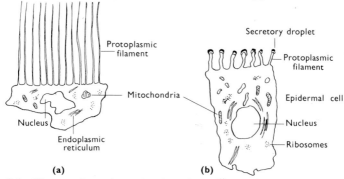

(a)   (b)

Fig. 2.4   Diagram of a resting and active epidermal cell. (a) Resting cell separated from the cuticle. (b) Active cell in the early stages of cuticle formation.

outer layer. On the inner surface it secretes a basement membrane which separates it from the haemocoele. The nucleus of the cell contains a small nucleolus; a few scattered filamentous mitochondria are present; only small amounts of ribonucleic acid occur. In the active state the cell is columnar or cubical in shape with easily distinguished cell boundaries; the nucleolus is large and the numbers of mitochondria and the amount of RNA have greatly increased (Fig. 2.4).

In all arthropods some of the epidermal cells become specialized for the production of secretions conveyed through ducts to the surface of the

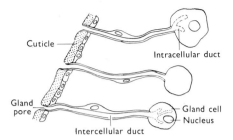

Fig. 2.5 Diagram of simple epidermal gland cell showing enlarged cell with intracellular duct separated from the epidermis. The products of the gland cell are conveyed to the exterior by an intercellular duct formed by another cell.

animal. These are the dermal glands (Fig. 2.5). They take no part in primary cuticle formation but sink below the other epidermal cells to lie in pockets projecting into the haemocoele. The products of these cells are very diverse, some producing defensive secretions, others sex attractants.

Many epidermal cells contain pigment granules but in some groups, e.g. Crustacea, specialized pigment cells, the chromatophores, occur (Fig. 2.6). These are usually large cells with elaborately branched processes

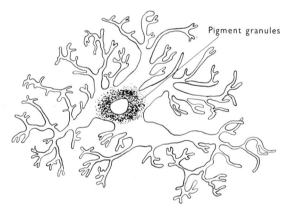

Fig. 2.6 Epidermal cell of a crustacean modified to form a chromatophore. The pigment granules, which may be either one of several colours (or granules of different colours may be present in the same cell) are shown concentrated around the nucleus. When they are dispersed through the distal ramifications of the cell, the cell takes on the colour of the dispersed pigment.

which ramify out amongst the other cells. A chromatophore may contain one or several pigments, common colours being red, yellow, black, white and blue. The pigment granules are capable of being moved; when bunched together in the cell body they are not readily visible to the naked eye but, when they are dispersed through the branches, the area takes on the visible colour of the pigment. When more than one pigment is present in a cell, each kind is capable of being dispersed or concentrated independently of the others.

The primary sense cells are derivatives of the epidermal cells and are modified to respond to mechanical or chemical or light stimuli by generating a series of electrical impulses which pass to the central nervous system. The cell body is produced distally into a sensitive process and proximally into a long axon which passes to the central nervous system.

The mechanoreceptors may have a single distal process associated with some cuticular filament or sheath (Type I neurons, Fig. 2.7a), or the process may be much branched and never associated with cuticular processes (Type II neurons, Fig. 2.7b). In Type I neurons, movement of the filament and sheath alters the electrical charge on the distal process, thus producing a generator potential. Electrical impulses in the axon vary in frequency with the size of this potential up to the limiting rate of the axon, and in insect mechanoreceptors the size varies as the generator potential changes. In Type II sense cells, the tips of these processes respond to mechanical strain and result in the generation of nerve impulses in much the same manner as the Type I neurons.

In the chemoreceptors (Fig. 2.7c) the distal processes are single or branched to give a small number of fine processes which may penetrate the cuticle wall through a tiny pore less than 1 μm in diameter. Each primary chemoreceptor is limited in the number of chemical substances that will stimulate it: for example, in the blow-fly one neuron is sensitive to sugars, one to monovalent salts and one simply to water. The total chemical sense of the animal depends upon the presence of a population of receptors, each reacting to a different substance. Chemoreceptors can be grouped under two classes, the contact receptors which are relatively insensitive and the olfactory receptors which are very sensitive indeed. Sensitivity to concentrations of less than 10^{-10} g/ml of the L-isomer of hexa-deca-dien-10–12-ol, a specific sex attractant, has been observed in *Bombyx*.

In photoreceptors the primary sense cell (Fig. 2.7e) is the retinula cell. The distal process is unbranched, forming a cylinder whose surface is covered with microtubules some 1000 Å long and from 400–1000 Å units in diameter. In primitive retinula cells the entire surface of the distal process is covered with microtubules, but in more advanced types the tubules are confined to one side and the nucleus and cell body extend distally (Fig. 2.7f).

The photosensitive pigment rhodopsin occurs in the lobster, *Homarus*. It is sensitive to a limited region of the electromagnetic spectrum. The entire population of retinula cells for all arthropods is sensitive only to the range 253–700 mμ.

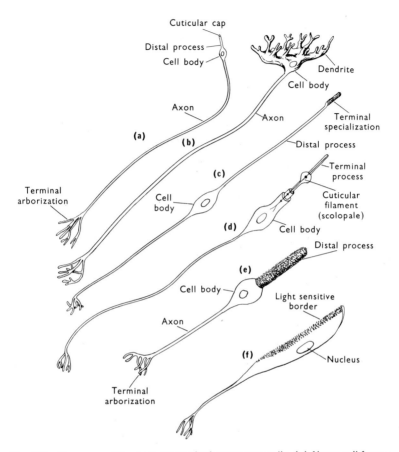

Fig. 2.7 Diagrams of the main types of primary sense cells. (**a**) Nerve cell from a simple mechanoreceptor, a Type I neuron associated with cuticular structures. (**b**) Nerve cell from a stretch receptor, a Type II neuron not associated with cuticular structures. (**c**) A nerve cell from a chemoreceptor. (**d**) A Type I neuron from a chordotonal sensillum showing a more elaborate cuticular structure, the scolopale. (**e**) A photoreceptor from the eye of a primitive arthropod showing a distinct distal process covered with microtubules. (**f**) Photoreceptor from a more advanced arthropod, the distal process and cell body having fused and the microtubules being confined to one edge of the cell.

The muscle cells

In arthropods unstriated or smooth muscle cells occur very infrequently. All the muscles of the Onychophora and the peduncle and mantle muscles

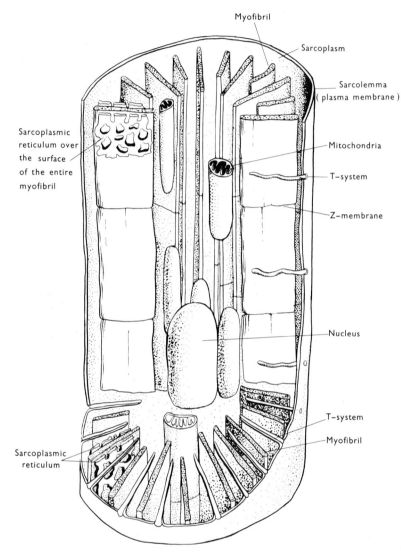

Fig. 2.8 Diagram showing the structure of a tubular muscle fibre. Only a part of the extensive sarcoplasmic reticulum, a system of spaces not opening to the exterior, is shown.

of the Cirripedia (barnacles) are of this type. A feature of smooth muscle cells is their ability to undergo considerable changes in length, an important characteristic where great changes in body shape occur, as in *Peripatus*. The muscles of all other arthropods are striated. This is correlated with the development of a skeletal system which restricts changes in body shape and of organ volume, and which, through a system of levers, allows appropriate magnification of small muscular movements. Under these conditions the muscles can produce their effects while undergoing minimal changes in length, giving the most efficient usage of the energy generated during a contraction.

All arthropod muscle fibres are multinucleate. Three types of fibre may be distinguished:

The tubular muscles (Fig. 2.8)

In these muscles there is a central core of sarcoplasm containing a row of nuclei. This core is surrounded by a radially arranged sheath of lamellar myofibrils separated by extensions of the sarcoplasm. Peripherally the plasma membrane encloses the muscle fibre. The sarcoplasm contains relatively few mitochondria and has a system of closed vesicles, the sarcoplasmic reticulum. The plasma membrane is also enfolded to form a system of cytoplasmic tubules (T-system) which opens to the surface.

The lamellar myofibrils are marked transversely by a number of bands which give them a striated appearance. These bands reflect the arrangement of the molecular apparatus which is the contractile machinery of the cell. In transverse section the electron microscope shows that each myosin filament is surrounded by six or more actin filaments in hexagonal arrangement.

In the fibres from the wing muscles of Orthoptera (grasshoppers), and Lepidoptera (butterflies), the radial pattern of the myofibrils has been lost and the central core of sarcoplasm is no longer present, the nuclei lying outside the myofibrils against the sarcolemma (Fig. 2.9). The myofibrils themselves are similar to those of tubular muscle but are closely packed together and tend to assume polygonal areas in cross-section. The mitochondria in the more advanced animals are large and regularly arranged: they are termed sarcosomes.

The fibrillar muscles (Fig. 2.10)

In the fibres from the wing muscles of the Coleoptera (beetles), Hymenoptera (bees, wasps) and Diptera (flies), the basic structure is similar to the above but the myofibrils are associated into small groups of $1-5$ μm in diameter which may be readily separated by gentle teasing. These fibrils give the name to the muscle type, fibrillar muscle. Within the sarcoplasm the sarcoplasmic reticulum is reduced to a large number of

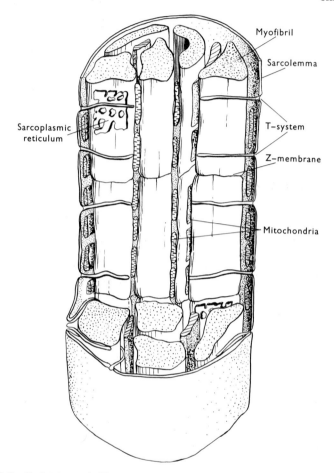

Fig. 2.9   Skeletal muscle fibres.

isolated vesicles and the cytoplasmic tubules (T-system) are confined to two places on the sarcomere. The mitochondria (sarcosomes) are large and regularly arranged opposite the A-bands.

The visceral muscles

The muscles of the visceral organs belong to the striated muscle type. However the alignment of the actin and myosin fibres is not necessarily uniform across the width of the muscle and a Z-disc may only occupy a few fibres, the others terminating in a Z-disc some further distance along the muscle fibres.

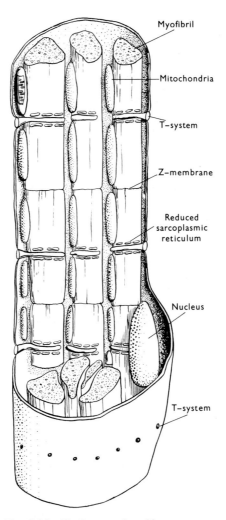

Myofibril

Mitochondria

T–system

Z–membrane

Reduced
sarcoplasmic
reticulum

Nucleus

T–system

Fig. 2.10 Fibrillar muscles of insects.

The secretory, absorptive, metabolic and excretory cells (Fig. 2.2)

In the cells so far considered, one of the many functions performed by primitive cells has become the dominant feature and there are imprinted upon the cell obvious specializations which serve its purpose. The other numerous functions of the cell are not absent: they still operate, though often at very low rates. Now we come to a group of cells in which the common metabolic responses of all cells, production of enzymes, uptake of nutrients, storage of reserve materials, excretion of waste and the synthesis of molecules, have developed to become dominant in their organization. In some cells all these functions may be exaggerated to some degree or one of them may become the specialized function of the cell. The majority of such cells are polarized. The uptake and output of materials is limited to their free ends at opposite sides of the cell. The organelles and vacuoles are concentrated nearer to one surface than the other. The uptake of materials through the cell surface represents the same problems of membrane permeability as are found in all cells.

Cells specialized for the production and secretion of enzymes

While all cells produce enzymes, normally the amounts produced are sufficient only for the cell's own needs. Cells which produce enzymes in excess of this and secrete the excess to promote chemical actions outside the cell body have a columnar shape, the nucleus set near the basal end, a well-developed Golgi apparatus, and globules of the specific secretion in the cell cytoplasm close to the secretory surface. The secretory surface is folded to produce a palisade of close-set projections giving a characteristic brush border. The material secreted either passes through the brush as droplets, leaving the cell and its border intact behind it, or it is discharged by the distal part of the cell breaking down and spewing the contents out. Cells of this type are found in the mid-gut, in the linings of the reproductive ducts, in the gonads and in the excretory organs.

Storage cells

The storage of reserve materials is a prominent feature of many types of body cells. Such cells are relatively large with a large nucleus and a prominent nucleolus. The cytoplasm in the fully developed cell is packed with granules of reserve material. The state of the cell is often related to the nutritional state of the animal. Many storage cells, such as those in the fat body of the insect, may store all classes of substances; some may store predominantly one or the other.

The majority of these storage cells absorb and secrete their products into the haemolymph, but in others where access to the haemolymph is limited, i.e. neurons, the storage cells may form a characteristic part of the tissue, being prominent and filled with food granules when the animal is well fed, shrunken and vacuolated when the animal is poorly nourished.

Special chemical cells

These are in many ways similar to the cells described above but studies have revealed that their function is the production of some chemical or some activity such as storage of uric acid. Such cells can only be characterized by identification of their particular chemistry. They are found through-out the arthropods and are diverse in function.

The cell arises from other cells by mitotic division; it then grows and specializes in its appropriate chemistry. Often this follows a series of changes, each of which may cause the cell to have a different appearance in histological preparations.

The endocrine cells

The chief endocrine cell in an arthropod is a neurosecretory cell (Fig.

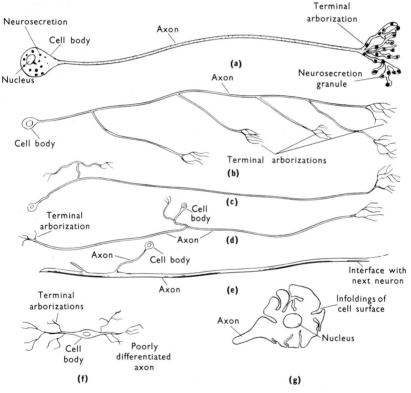

Fig. 2.11 Diagrams of some types of nerve cells found in arthropods. (**a**) A neuro-endocrine cell; neurosecretion is formed in the cell body and moves down the axon to accumulate in the terminal arborizations from which it is eventually released. (**b**) and (**c**) Two motor neurons showing different patterns of axon form. (**d**) A bipolar neuron. (**e**) A giant fibre neuron. (**f**) An amacrine neuron, an association neuron lacking well-defined axons. (**g**) Cell body of an insect neuron showing the deeply indented surface into which folds of other cells penetrate.

2.11a). Cytologically, neurosecretory and neuro-endocrine cells are very similar. A neurosecretory cell should only be called a neuro-endocrine cell when its products have been shown experimentally to have an endocrine function.

The other endocrine cells in arthropods are found in the prothoracic glands and the corpora allata of insects and in the androgenic glands (sex glands) of certain Crustacea.

In the prothoracic gland the cells which secrete the hormone ecdysone are large and contain a large nucleus which, in the moth *Cecropia*, is complexly folded; a prominent nucleolus is present.

In the cells of the corpora allata which secrete the juvenile hormone, the nucleus is again large and more or less lobulated with a single nucleolus. The endoplasmic reticulum is well developed; it is often agranular, indicative of steroid hormone synthesis, but in some cases it is rough (granular), indicative of protein synthesis. Granules are present in the cytoplasm, first close to the nucleus and later becoming more evenly distributed. Giant polyploid cells occur which, except for size, appear identical with normal corpora allata cells.

In the androgenic glands the nuclei are peripheral in location and the cytoplasm vacuolated.

The nerve cells (Fig. 2.11)

The neurons are those cells of the nervous system which are specialized for the rapid conduction of discrete electrical impulses. Each neuron consists of a cell body, the perikaryon, from which long cytoplasmic processes, the axons, radiate out towards other cells. The axons end in small fine branches, terminal arborizations, which make contact with other cells. Terminal arborizations may also arise along the length of the axon as well as at its termination.

The perikaryon may consist of either a large nucleus showing very little chromatin and surrounded by abundant cytoplasm, or a small nucleus with dense chromatin, surrounded by a thin layer of cytoplasm. The cytoplasm contains a well-developed endoplasmic reticulum and Golgi apparatus. Granules of RNA are present, normally scattered throughout the cytoplasm, but when damaged, or under circumstances of increased protein synthesis, the granules aggregate together and resemble the Nissl substance of vertebrate neurons. The plasma membrane of the cell body may be deeply indented, forming crevices that may sometimes penetrate to the nucleus. The ability of the neuron to conduct electrical impulses is largely confined to the axon. Transmission of pulsed changes in permeability along the axon allow momentary changes in the distribution of the ions which appear as a change in electrical potential. In the time taken to re-establish the distribution of ions following a pulse, the nerve is refractory to further transmission.

At certain specialized places, the synapses, the permeability pulse or series of pulses results in the liberation of a chemical 'transmitter' substance from clusters of minute synaptic vesicles (300 Å in diameter) that lie in the cytoplasm just beneath the plasma membrane (Fig. 2.12). Where the

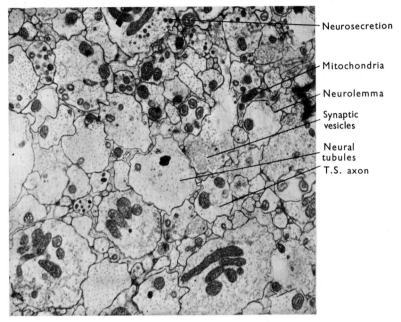

Neurosecretion

Mitochondria

Neurolemma

Synaptic vesicles

Neural tubules

T.S. axon

Fig. 2.12 Section through the neuropile of the frontal ganglion of a locust showing the axons containing mitochondria, neurosecretion, neural tubules and synaptic vesicles. (Courtesy of R. Allum)

synapse is between neurons the transmitter substance is acetyl-choline but where the contact is between neurons and other types of cell other substances occur. Glutamate and GABA (gamma-amino-butyric-acid) have been identified in neuro-muscular synapses.

In determining the role a neuron plays in the collecting and processing of information within the nervous system, its geometric form and dendritic pattern is of paramount importance. Great differences occur between neurons: some of the forms which they assume are indicated in Fig. 2.11. A neuron may be passive and respond only when stimulated to do so; many neurons, however, are 'spontaneously' active. That is, they continuously generate patterns of impulses which originate from endogenous processes within the cell.

The tracheolar cells (Fig. 2.13)

These cells, which are widespread throughout terrestrial arthropods, are associated with the conduction of air into the immediate vicinity of the respiring cell. The tracheolar cell sends a short process to the trachea and,

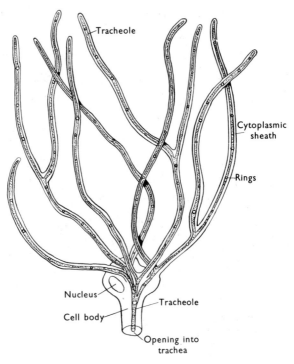

Fig. 2.13 Diagram of a tracheolar cell, the tracheoles are shaded and the rings indicate structures that help to keep the lumen of the tracheole open. In general the tracheoles should be much longer in proportion to their width as indicated in the diagram, the cytoplasmic sheath very much thinner, and the rings are usually only seen in E.M. examinations.

distally, several very long and thin processes which spread out and lie over the metabolizing cells. Within these processes are fine intracellular tubules $0.25 - 0.30$ μm in diameter, held open by rings which occur at intervals along their length. In the vicinity of the cell body they open into a common tube which opens to the air. The tubes end blindly near the tips of the processes. In a resting condition the tubes are full of fluid; in an active condition the fluid is withdrawn and air sucked in.

The haemocytes

These are individual cells which originate from the mesoderm and circulate within the haemolymph and/or occur loosely adherent to the basement membrane of epithelia or amongst connective tissues. Only three of the very many known types of haemocytes will be mentioned here:

1. A large granular-containing cell with a single large centrally located nucleus; the cell is amoeboid and may be the principal phagocyte cell of the haemolymph. Examples are the plasmatocytes of Insecta and the granular cells of Crustacea.
2. A small agranular cell with a large round nucleus and a small amount of cytoplasm. Examples are the agranular cells of Crustacea and prohaemocytes of Insecta. These cells are thought to be the precursors of the other types of haemocytes circulating in the blood.
3. Fragile cells with a small round cartwheel-like nucleus. These cells, which disintegrate quickly, are the coagulocytes.

Many of the haemocytes do not spend their entire existence floating freely in the haemolymph but become stationary for periods of time, adhering to the walls of the haemocoele. The cells are versatile in the shape they assume: the same cell will present a very different picture when rounded and floating in fluid from the one it presents when adhering by one surface to the haemocoele wall with the other stressed by the passing flow of haemolymph.

Following injury or wounding, the haemocytes are the cells which migrate to the area and seal off the wound; secretions from them initiate clotting of the haemolymph at the wound surface.

Many of the haemocytes contain within their cytoplasm nutritive reserves which may be in transport from one site to another and some appear to act in the transport of specific substances, hormones, from one place to another, or may in fact produce specific secretions that are released into the haemolymph or adjacent to their site of action.

3

The Tissues of Arthropods

Cells are associated together in the animal body to form tissues which have characteristic structures for definite functions. Within each tissue one or a few cell types carry out the special function while other cell types 'service' them. Based upon the arrangement of the major cell type, three main kinds of tissue can be recognized:

1. Tissues forming a sheet of epithelium one cell thick (Fig. 3.1a) e.g. epidermis.
2. Tissues in which the cells have a three-dimensional array (Fig. 3.1b) e.g. nerves, ganglia.
3. 'Open tissues' in which the cells are separated by considerable amounts of intercellular material (Fig. 3.1c).

The services provided by the 'service' cell types comprise support, given by cells such as connective tissue and tracheolar cells; movement, produced by muscle fibres; the supply of nutrients and the removal of waste by the haemolymph and haemocytes, often conducted to the main cells by blood vessels; the supply of oxygen and removal of carbon dioxide either via pigments in the haemolymph or by the tracheolar cells; and the integration and control of the tissue activity by neurons.

THE EPITHELIAL TISSUES

The cellular sheet characteristic of these tissues consists of polygonal, cubical or columnar cells each in close contact with its neighbour, with one surface set on a basement membrane. This membrane may be secreted by

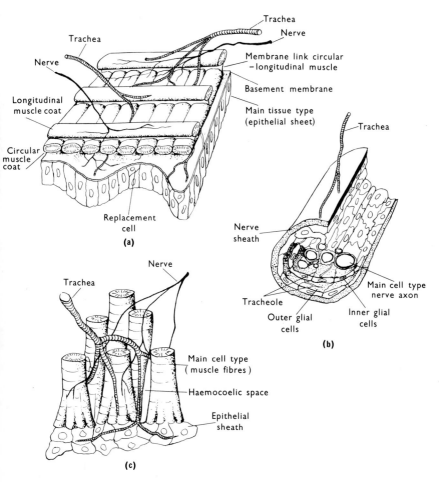

Fig. 3.1 Diagrams showing the three main classes of tissue based on the arrangement of the major cell type. (**a**) Epithelial tissues (gut). (**b**) Three-dimensional
array (nerve). (**c**) Open tissue (muscle).

the cells themselves but sometimes it is the product of haemocytes adherent
to its surface.

Areas of special contact, the desmosomes, occur between the plasmalemmas of adjacent cells and hold the cells closely together (Plate 1). They are
absent in insect fat body where the cells undergo changes in volume. In
cell sheets where much activity is occurring the cell boundaries are
distinct but, where the cells are inactive and shrunken, boundaries are

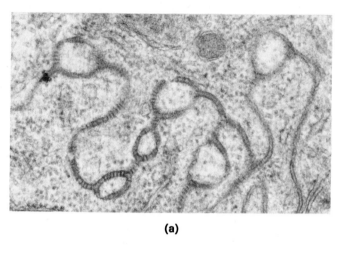

(a)

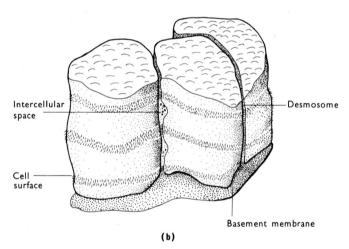

Intercellular
space ——————— Desmosome

Cell ———
surface

Basement membrane

(b)

Plate 1 (a) E. M. photograph of septate desmosomes from the recurrent nerve
of a locust (Courtesy R. Allum). (b) Diagram of epithelial cells indicating the
probable bandlike nature of the desmosome on a cell surface.

either absent or very hard to demonstrate. The life span of the cells composing the sheet is shorter than that of the sheet itself. The dead cells are
ejected from the sheet and are replaced by division from unspecialized cells
scattered in pockets amongst the main sequence cells.

The cells of the epithelium may not all be in the same phase. Their coordination from random activity into the same phase is one of the ways in which epithelial tissues attain their peak activity.

In many tissues the sheet of cells is flat, or curved to form a simple cylinder, but in others the sheet may be folded in a complex way. When the sheet is folded it often forms pits or pockets in its surface, and replacement cells tend to aggregate at the bottom of these pits. The surface of the epithelium is then composed almost entirely of fully differentiated main-sequence cells.

In epithelial tissues the service cells lie on the side of the basement membrane away from the epithelium. The muscles which suspend the epithelial sheet consist of fibres whose axes are normally perpendicular to the sheet, their myofibrils being inserted into the basement membrane. Those muscles which move the sheet itself have their long axes parallel to the surface of the basement membrane, to which they are fastened by connective tissue bundles. The muscle fibres may form an irregular network over the surface where diffuse or general movement of the sheet is required, but where regular and frequent movements occur the muscle fibres are arranged in one or more sheets with their axes at right angles to one another. Such muscle sheets are usually one fibre thick. Where the epithelial sheet forms a cylinder of small diameter a single muscle fibre may spiral round the tube.

The tissue cells are separated from the haemolymph by a thin layer of cells, the endothelium. They enfold the entire surface of the tissue, penetrating into cracks and crevices. Blood may be conducted close to the tissue by arterioles which open out into spaces where the haemolymph actually makes contact with the tissue.

Nerves penetrate into the tissues, the terminal arborizations coming into synaptic connection with the tissue cells. Both excitory and inhibitory control functions are exercised by the neurons innervating the muscle fibres.

In the majority of epithelia the main features of the tissues are retained throughout their existence. The phases of formation, growth, function and death of the cells that make the tissue up tend to be randomly distributed amongst them, so that with the passage of time the overall picture of the tissue is little altered. The main changes are reflections of the ebb and flow of supply and demand. In the epidermis and the cuticle however, cyclical changes of long period occur, which so dominate the tissue that a description of it at any one time cannot do duty as a basic account, nor are its structure and function intelligible without reference to the time element.

The cycle generates through the interaction of growth in tissue weight and volume, and the presence of an inelastic cuticle covering the outer surface of the animal. Under these conditions the maintenance of growth

and of cuticular function are possible only if the following sequence of changes occurs (Fig. 3.2).

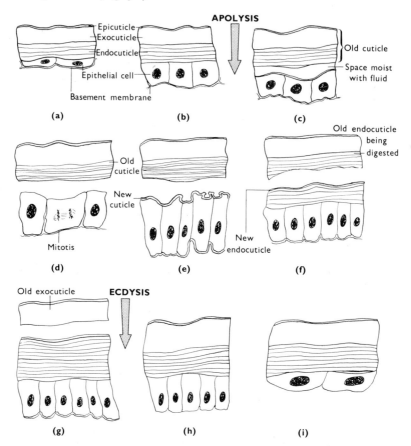

Fig. 3.2 Diagram of the sequence of epidermis and cuticle changes throughout a moulting cycle. (**a**) Resting epidermis. (**b**) Cells enlarging and becoming active. (**c**) Apolysis, separation of cells from the old cuticle. (**d**) Mitosis. (**e**) Commencement of formation of the new cuticle, secretion of the epicuticle. (**f**) Digestion and absorption of the old endocuticle and secretion of the new procuticle. (**g**) Old endocuticle completely digested away, new procuticle increased in thickness. (**h**) The old cuticle has been shed by ecdysis. (**i**) The epidermal cells have returned to a resting state.

1. The epidermis separates from the old cuticle (apolysis) and grows by cellular multiplication, giving a greatly extended epidermis much folded under the old cuticle.

2. The new cuticle is secreted upon the surface of the epithelium.
3. The old cuticle is shed: this is the actual moult or ecdysis.
4. The epidermis and the new cuticle are extended to their maximum size and maintained there until chemical processes harden the new cuticle and lock it in its new dimensions. The space thus created becomes available for further tissue growth.
5. Following this the cuticle increases further in thickness, the epidermis remaining active.
6. The cycle (called the moulting cycle) is repeated when the tissues have again filled the available space, or terminates in phase 4 at cessation of growth in the adult instar.

In some arthropods the cuticle is further hardened and strengthened by the secretion of calcium salts into it; these salts are often absorbed at moulting and are then used again for the new cuticle.

In terrestrial arthropods water loss through the cuticle is reduced or prevented altogether by the secretion from the epidermal glands of a wax layer outside the cuticle. Also, in terrestrial animals, water uptake may occur from humid atmospheres. This is possible because of small pores 100–200 Å in diameter which occur in the cuticle and in which water vapour can condense out on their walls from atmospheres with a relative humidity of 55% or over.

The surface of the cuticle frequently exhibits complex patterns of folds, ridges, hexagons and other shapes (Plate 2). These are formed by the cuticle being secreted at different rates from the surfaces of the cells and from the stress that these cells are subjected to during cuticle formation, i.e. the manner in which they are folded. The various points at which the first cuticle layer appears are determined by the structure of the cytoskeleton occupying the cell between the nucleus and the plasma membrane, these points thereafter secreting more actively than the rest of the cell surface.

THREE-DIMENSIONAL TISSUE ARRAYS

In these tissues, where the main cell type assumes a three-dimensional array, the main problems are the maintenance of cells in their spatial relationships, the conveyance of nutrients and metabolic products to and from the cells, and the growth of the tissue by cellular multiplication. In arthropods, the main three-dimensional tissues are those of the brain and the ganglia. The actual volumes of these tissues are small, probably because the animals lack adequate means of circulating haemolymph through such closely knit cell groups.

THE NERVE TISSUES

The position of each neuron, its shape and its specific contacts with

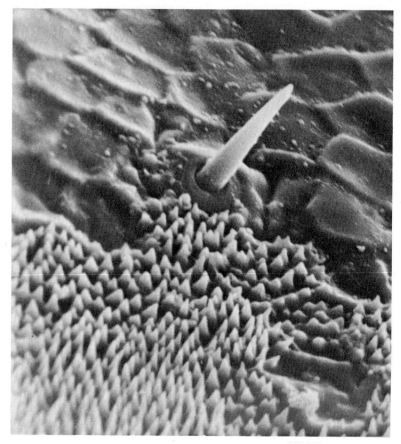

Plate 2 Surface of the cuticle from the lateral posterior region of the head of a locust. The scale like surface, probably the primitive cuticular pattern, is sharply replaced by another surface pattern more posteriorly. A sensory hair, probably a mechanoreceptor, projecting from its socket is shown in the centre of the picture.

other neurons dictate the basic information-processing properties of the system and these must therefore be rigidly fixed. The positions of the cell body and the major axon pathways once laid down do not change.

Between neurons occur points of contact, the synapses, through which the activity of one neuron affects the next. Synapses occur mainly between terminal branches of the axons, occasionally between the axons themselves. Only in a few cases, as in some neurons in the brain of the crab, does the synapse occur on the perikaryon. Each pre-synaptic axon makes many post-synaptic contacts: two pre-synaptic axons may synapse with the same post-synaptic axon and serial synapses may also be present. In the majority of synapses the neurons are separated by a space some 200–300 Å units wide. The arrival of a train of impulses results in the release of a transmitter substance into this space; when enough has been released, electrical activity is generated in the post-synaptic neuron. The majority of synapses permit conduction in only one direction. In some synapses however, the pre- and post-synaptic membranes are in very close contact and the electrical pulse may pass from one to the other without chemical intermediaries.

The passage of the pulse down an axon requires ion (Na^{2+}) exchange between the axon and its surrounding medium. The axon is therefore surrounded by a liquid ion-containing fluid for its proper function and is not in actual contact with another cell.

The neurons, their axons and terminalizations, except where synapses occur, are sheathed by glial cells which also tend to form an outer cell layer around the ganglia and major nerves.

Glial cells (Fig. 3.3) closely enfold the perikaryon whose surface is indented, processes of the glial cells penetrating into these cracks and crevices. Thin films of cytoplasm from the glial cells, sometimes only 250 Å in width, penetrate and fold around the axons, always, however, separated from them by intercellular spaces. These spaces are maintained by processes called septate desmosomes stretching between the plasmalemmas of the glial cell and neuron. One glial cell may ensheathe a single axon if this is of sufficient size, but will ensheathe 20–30 axons of small diameter. Only in a few instances, e.g. the crab brain or the ventral cord of some decapod Crustacea, does the glial cell make turns around the axon to produce a myelin sheath with nodes of Ranvier, similar to that characteristic of vertebrate axons.

Outside these glial cells which are so intimately associated with the neurons there is another layer of glial cells (Fig. 3.3), also knit together by extensive folding and having a similar association with the inner glial cells. The cells in this outer layer are very rich in mitochondria and contain granules of glycogen and droplets of lipid. Between the cells there is an extensive lacuna containing fluid continuous with the much smaller spaces around the individual axons.

The glial cells serve to support the neurons and also to convey nutrients to and metabolic products away from them. Stores of material are held in the outer glial cells of well-fed animals but only the vacuoles are visible in those of starved arthropods.

The outer glial cells are separated from the haemolymph by an extracellular sheath consisting of a meshwork of collagen-like fibres in a finely granular (muco-polysaccharide) matrix. This sheath encloses the entire nervous system, but is markedly thicker over the ganglia than it is where it invests the nerves.

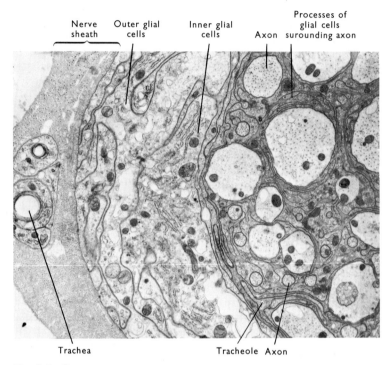

Fig. 3.3 Transverse section of a nerve from the frontal ganglion of a locust. On the left is the nerve sheath penetrated by two small trachea; on the right of the sheath are the outer glial cells; to the right of these the inner glial cells penetrating amongst the nerve axons on the right of the picture. (Courtesy R. Allum)

Scattered throughout the nervous system are the neurosecretory (endocrine) cells. These have much the same relationships with the glial cells as do the ordinary neurons. The neurosecretory material in some cases passes through the cell wall into the spaces between the neurons and the glial cells. In other cases the cell terminalizations end in an organ, the neurohaemal organ, where release of the material into the haemolymph occurs. The neurohaemal organ may simply consist of the swollen ends of the axons containing neurosecretory material imbedded in glial cells separated from the haemolymph by a thin neurolemma, or it may also possess other types of cells to give a much more elaborate structure.

Once the initial number of neurons is established when the cells are differentiated in the embryo, there is little or no increase in their number in the subsequent growth and development of the arthropod. Growth of the nerve tissue is brought about by the following changes:

1. Increase in size of the neuron bodies occurs. Fully grown neurons may have up to 70 times the volume of the initial cells.
2. Sensory neurons differentiate at the periphery to increase or maintain the density of sense organs in the integument and these send back axons which enter the ganglia. For example, the number of hair receptor axons to the abdominal cord of *Acheta domestica* (cricket) increases from 50 to 750 during post-embryonic growth.
3. There is an increase in the number and the size of the glial cells. In the cricket the increase is from 700 to 10 000 during post-embryonic growth of the animal.

THE OPEN TISSUES

The main cell types of these tissues are the haemocytes and the muscle fibres. These cells do not form a close-knit array but are separated sufficiently for the haemolymph to circulate between them.

THE MUSCLE TISSUES

Muscle tissues are formed from an array of muscle fibres, each fibre being limited by its sarcolemma. Small muscles may consist of only a single fibre. Larger masses of muscle tissue comprise either a few greatly enlarged fibres, as in the wing muscles of insects, or a large number of smaller fibres, as in the femoral leg muscles of grasshoppers. The reasons for associating these fibres together as a single muscle are as follows: their anatomical origins and insertions indicate their unitary action; the different properties of each fibre confer different abilities on the action of the muscle; the neurons servicing them link together so that they respond as a unit. The proportions of different fibre types confer upon the muscle characteristic properties closely adapted to the habits of the animal.

The groups of muscle fibres are further knit into a tissue organization by the distribution of neurons amongst them. In many muscle tissues a single neuron will innervate all the fibres present while another neuron passes to only a few of them. For example, in the leg muscles of the rock lobster, 38% of the fibres are innervated from one neuron, 26% from two, 29% from three and 7% from four neurons. One neuron is distributed to 90% of the muscle fibres present.

THE BLOOD TISSUES

Blood consists of a fluid of considerable chemical complexity, in which a number of different types of cells may be found. 'Service' cells can hardly be said to occur in this tissue. All its cells are better regarded as main cell types.

One of the characteristics of this tissue is its extreme variability in composition, both in time in any one individual, and between individuals. The volume and chemical content of the fluid part of the haemolymph varies both with the cyclical events that characterize the growth of the animal and with the nutritive state of the creature. The respiratory pigments, haemoglobin or haemocyanin, when present are in solution in the haemolymph and not contained within the haemocytes.

The number of cells present is also a variable quantity. In large arthropods an average number of 75 000 per mm^3 is recorded from both Crustacea and Insecta but in small arthropods much lower numbers are present. Haemocytes undergoing mitotic division while circulating in the haemolymph only occur when there is stress demanding all the available cells present.

In addition to the haemocytes, many other cell types may occasionally be found in the blood, especially where the animal is undergoing metamorphosis and when organ formation and/or dissolution are occurring.

4

The Organs of Arthropods

INTRODUCTION

The organs of arthropods are distinct and easily recognizable anatomical entities specialized for the performance of some body function or functions. Organs are composed of a number of different tissues arranged according to some structural design for each organ. This design creates new properties and sets new limitations in addition to those possessed by its component tissues. A study of the organs of a wide range of arthropods shows that the basic design of any one organ type is detectable in all its members. Descriptions are given here of the basic design and major modifications for both structure and function of the organs of arthropods. Relative constancy at the level of organ design does not necessarily extend to the tissues and cells which compose it. These may change independently of organ design thus conferring new functions or losing old ones. Changes may occur in organ design which are not reflected in tissues and cells whose functions then remain unchanged.

The principles of organ design are more clearly revealed when organs are classified according to the differences and similarities of their structure and not according to the body systems of which they are a part. In arthropods the following classification is appropriate:

1. The appendages.
2. The integumental organs.
3. The tubular organs.
4. The brain and ganglia of the nervous system.
5. The diffuse organs.

THE APPENDAGES

The appendages of arthropods are paired ventro-lateral out-pushings of the body wall (Fig. 4.1). In living arthropods they occur in their simplest form in *Peripatus*. The surface of the appendage is covered with cuticle beneath which is an epidermis resting upon a thick basement membrane. The circular and longitudinal muscles of the body wall form the intrinsic musculature of the appendage; an extension of the haemocoele occupies the

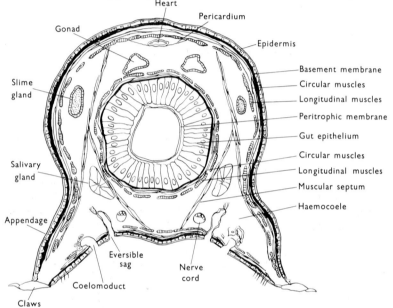

Fig. 4.1   Diagram of a transverse section through the trunk of *Peripatus*.

lumen of the appendage. The appendages are moved relative to the body wall by their extrinsic muscles arising from the appendage and inserted on the body wall. These muscles produce an anterior promotor and posterior remotor swing of the appendage with some dorso-ventral movement, so that the tip of the appendage describes an elongate oval movement relative to its base. There is some rotation of the appendage about its long axis. In the absence of a rigid skeletal support the appendage maintains its shape by an interaction between the hydrostatic pressure of its core which distends it and the contraction of its muscles which bring about movements and changes of shape in opposition to this.

There are two main types of appendage in *Peripatus*: the pre-antennae and the trunk limbs (Fig. 4.2a,d). The pre-antennae arise from the first

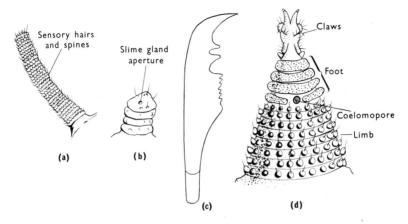

Fig. 4.2 The appendages of *Peripatus*. (**a**) The pre-antenna, (**b**) the slime papilla, (**c**) the mandible, (**d**) a trunk limb. In this animal the cuticle does not act as an exoskeleton, having a protective function only.

body segment and are mobile elongate cylindrical feelers. Each is ringed with sensory hairs and spines and capped at its tip with a pad of spiniferous tissue. A series of short muscular cylinders down its length give it mobility. The trunk limbs are cone-shaped; on the ventral surface towards the distal tip of the cone are three pads upon which the animal normally walks; distally there is a sclerotized foot ending in a pair of claws; the feet and claws can be retracted into the proximal part of the limb. They are only used in locomotion when the animal is in danger of slipping. The limbs of the two segments immediately following the antennae are modified: those of the second segment are sunk into the buccal cavity on each side of the mouth. The proximal part of the limb supplies the muscles which operate the foot and claws, which now have the form of a sclerotized rod with a pair of large teeth at its distal end (Fig. 4.1c). The jaws move anteriorly and posteriorly and rasp away the flesh of the animal prey. The second pair form the slime papillae (Fig. 4.1b); the limb base is retained but the foot is lost. At the tip of the limb cone is a pore through which the animal can forcibly eject a slim pencil of slime, the mobile papillae permitting it to be shot in any direction. This slime is used for defence purposes only.

In all other arthropods the cuticle forms an exoskeleton and this profoundly affects the design of the segmental appendages (Fig. 4.3). Their basic form is an elongate slender cylinder, similar in structure to that described above, but now ensheathed in a series of sclerotized cylinders, the podites or podomeres, between which there are shorter lengths of non-sclerotized cuticle forming flexible annuli. Extrinsic muscles arising from the body wall may be inserted into the podomeres. Muscles arising from the proximal podomeres are inserted into more distal ones, thus conferring considerable mobility on this articulated appendage. The flexible annulus between the sclerotized cylinders permits the more distal podite to be moved towards any point in a cone relative to the proximal one (Fig. 4.3).

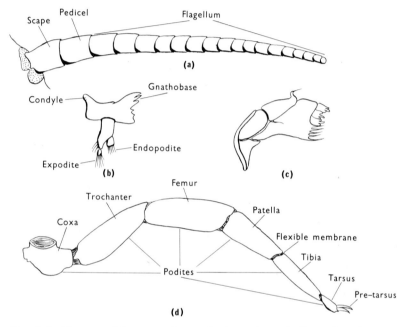

Fig. 4.3 Diagram of the appendages of other arthropods illustrating changes associated with the development of an exoskeleton. (**a**) Antenna from a myriapod. (**b**) Mandible from a crustacean (Copepoda). (**c**) Mandible from an insect. (**d**) Trunk limb from a scorpion.

To achieve this and to hold the podomeres in any given position requires an extensive musculature, considerable neural integration and continuous expenditure of energy. Only where unrestricted movement is important and mechanical requirement sufficiently slight is this situation found. More often adaptive radiation of the appendage and the increased efficiency demanded from it, have led to restricted mobility, increased strength, and development of precision of movement of which the simple basic design is not capable. To confer these properties, the articulations between the podomeres and their relationships to each other are as follows:

1. Complete rigidity is achieved by podomeres fusing together; alternatively, even greater flexibility, though with increased weakness, can be

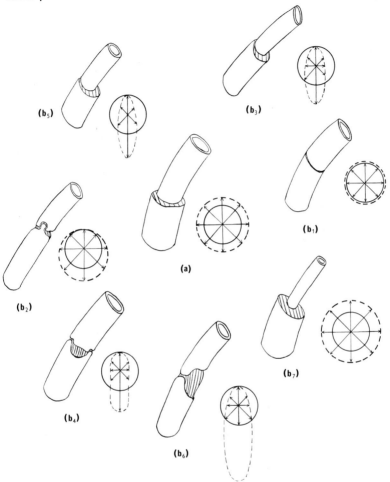

Fig. 4.4 Diagrams to show the main types of podite articulation. (a) Simple membrane annulus allowing moderate movement in any direction. (b_1) Annulus absent, completely rigid joint. (b_2) Single ball and socket joint giving mobility with strength. (b_3) Annulus shorter at one place than at others giving restricted mobility with little strength. (b_4) Two ball and socket joints restricting movement to one plane and giving great strength. (b_5) Annulus shortened, to give a hinge around part of the circumference, mobility not so rigidly restricted to one plane as above, moderate strength. (b_6) Amplitude of movement in one plane greatly increased by emargination of the podites while still allowing a hinge articulation. (b_7) Increase in amplitude of movement in every direction by the distal podite being of smaller diameter than the proximal one, little strength. The figures beside the diagrams indicate the direction and range of movement of the distal podite relative to the proximal one. The inner circle represents the proximal podite, the dotted one the possible movements of the distal one.

achieved by the distal podomere being smaller in diameter than the proximal one (Fig. 4.4a, 4.4b$_1$, 4.4b$_7$).

2. Much mobility can be retained with greatly increased strength by the development of a ball and socket joint at some point of the perimeter of the podomeres (Fig. 4.4b$_2$). This still requires extensive musculature, neural integration and energy expenditure for its efficient working.

3. Restriction of movement with slight increase in strength occurs when the membranous annulus, instead of being of equal length all round, is shorter on one side than elsewhere (Fig. 4.4b$_3$). This limits the main movement to a plane at right angles to the place of shortening, allowing an economy of muscles since there is no need for a complete muscle cylinder to be present. Some movement in planes other than the main one is possible.

4. The development of two points of articulation restricts the limb movement to a plane at right angles to a line joining the articulations (Fig. 4.4b$_4$). Movement in this plane is increased by emargination of the edge of the podite in the plane of movement. This type of joint can allow a wide angle of movement on either side of the hinge axis while maintaining great strength.

5. Close contact may occur between a considerable length of the edges of the podomeres, forming a true hinge (Fig. 4.4b$_5$). These articulations are strong and may have a great range of movement in a plane at right angles to the hinge but only on the side away from the hinge, which may be emarginated. Bending of the hinge backwards is often accomplished by pressure increases in the haemocoelic core of the limb.

The cylindrical exoskeletal structure of the limb confers considerable strength for a minimum amount of material. It limits the amount of muscle that can be contained within the hollow of the cylinder. Some increase in muscle size can be accommodated by increasing the diameter of the cylinder, but soon the lightness and strength of this kind of construction are lost in supplying the vastly greater mechanical strength such an increase demands. There is as little weight as possible in the more distal podomeres, since weight here requires a greater skeletal bulk and musculature of the more basal podomeres to hold and move the limb. Muscles are therefore concentrated in the basal podomeres, the more distal ones being operated from them by tendons passing down the limb.

The core of the limb is a long slender extension of the haemocoele and contains haemolymph. To ensure circulation of blood through the limb there is a slender artery conveying blood almost to the limb tip, the return flow occurring through the haemocoelic space surrounding the artery. Where the arterial system is virtually non-existent, small accessory pumps occur at the limb base. Alternatively, a membrane is present which divides the limb haemocoele almost to the tip. Near the base of the limb, muscles are present and cause the membrane to oscillate, the oscillations driving the blood up the limb on one side, aiding its return flow on the other. The membrane appears to be present in appendages of considerable diameter and the pulsatile vesicles in those of narrow diameter.

The main appendage types in arthropods

A study of the numerous types of appendages found in arthropods possessing an exoskeleton indicates that they are modifications of three basic designs:

The uniramous appendages (Fig. 4.5a)

These are the appendages of the Myriapoda and Insecta which consist of a single series of podomeres. Movable lobes, the endites, may occur on the inner side of the basal podomeres and exites on the outer edge in a few instances. Their presence is not a widespread feature of this type of appendage.

The biramous appendages (Fig. 4.5b)

These are the appendages of the Crustacea; there are two podomeres basally forming the protopodite. From its distal end arise two rami, the exopodite and the endopodite. These may be of equal size but often the endopodite forms the main shaft of the appendage and the exopodite a slender flexible ramus.

The trilobite—arachnid appendage (Fig. 4.5c)

In the Trilobita there is a main series of podomeres; but from the basal podomere on its dorsal side there arises a long jointed setae-fringed exite, the epipodite. The appendage limb has a biramous structure but the origin of the second ramus is different from that of the Crustacean appendage.

In the Merostomata a basal epipodite occurs on the proximal podomere of the main shaft of the appendage, indicating relationships to the Trilobite type of appendage. The evolutionary relationships between the Merostomata and the Arachnida link the appendage structure of the latter with that of the former although no epipodite occurs on any arachnid appendage. The similarity in uniramous structure between the appendages of Myriapoda-Insecta and Arachnida is due to convergence and not phylogenetic relationship.

The adaptations of arthropod appendages

The appendages have become specialized to perform a variety of actions such as locomotion, grasping and handling, mastication, and food gathering, and can function as sensory feelers and copulatory organs. These have

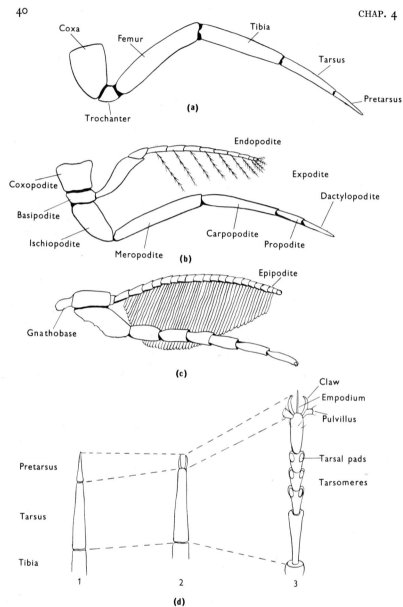

Fig. 4.5 The main appendage types in arthropods. (**a**) Uniramous limb. (**b**) Biramous limb. (**c**) 'Trilobite' limb. (**d**) Differences in the limb tips: **1** simple limb tip, **2** a simple pair of claws, **3** limb tip of an insect showing claws, empodium, pulvillus, and sub-division of the tarsus into tarsomeres.

influenced the design of the appendage so that, no matter what its basic pattern or wherever it occurs, its functions can be recognized from its structure. With the exception of the sensory feelers of the first segment of the body, all appendages developed some locomotory ability before specializing for other functions.

The walking limbs of myriapods and insects are an example of a purely locomotory appendage. The limbs are approximately S-shaped, consisting of a single series of podomeres called in sequence from the base distally: coxa, trochanter, femur, tibia, tarsus and pre-tarsus. The coxa is directed ventro-laterally and is articulated to the body wall so that it swings anteriorly and posteriorly (promotor and remotor movements). The trochanter is articulated to the coxa dorsally and ventrally and moves anteriorly and posteriorly in a horizontal plane. The trochanter-femur articulation permits movement dorsally and ventrally, as does the femur-tibia joint. This articulation forms the knee joint so that the movements of the tip of the tibia tend to be towards and from the mid-line of the body. The tarsus and pre-tarsus articulate so that movement is in the same plane as that of the tibia. The S-shape of the limb, however, again means that these articulations tend to bring the tip of the appendage close to, or further away from, its base.

In many instances the coxa develops a second point of articulation with the body wall, thus rigidly confining its movement to a swing at right angles to a line joining these articulations. This also abolishes the rotary movement about the long axis of the limb. Almost all pterygote insects show this modification.

In the Myriapoda and more primitive Insecta the pre-tarsus tapers to a point, the tip of which makes actual contact with the ground. From the tarsus and pre-tarsus develop more elaborate structures which aid the animal to grip or move on substrates where point contact is inadequate (Fig. 4.5d). The pre-tarsus forms a pair of retractile claws articulated to a basal plate, which also bears laterally a pair of pad-like projections, the pulvilli, and central bristle, the empodium. The tarsus becomes subdivided into a number of tarsomeres, articulated to provide considerable flexibility at the tip of the limb. Muscles do not occur between the successive tarsomeres. Contact with the ground is through pads developed on the underside of the tarsomeres and through the tip of the tibia which may bear strong spines. The pulvilli and empodium may develop into pads which permit the animal to cling to smooth surfaces. These modifications greatly facilitate the animal's ability to move freely on very different types of substrate.

In the walking limbs of Crustacea the main shaft of the limb is formed from the protopodite and the endopodite. The podomeres in sequence from the base outwards are called: coxopodite, basipodite, ischiopodite, meropodite, carpopodite, propodite and dactylopodite respectively. The limb is also S-shaped, the sequence of articulations being similar to that already described. The tip of the limb is never elaborated as it is in insects to form the tarsus and pre-tarsus. The exopodite, which originates from the distal end of the basipodite, is an elongate many-jointed ramus fringed with hairs.

Besides propelling the crustacean forwards, the movements of the limbs, especially those of the exopodites, create a complex of water currents about the limbs and body of the animal. Detritus, some of which has food value, is caught up in the water movement and carried by this complex of currents towards the mid-ventral line of the animal. Here, a forward directed current carries it to the mouth, where manipulation by the base of the limbs and the action of the broad labral plate lying in front of the mouth direct it into the gut. The trunk limbs of Crustacea thus have a dual function of locomotion and food gathering, which, primitively, were probably of equal importance.

The locomotory function and structure of the limbs of trilobites appear similar to those of Crustacea and the epipodite of these limbs may have had a similar function to that of the exopodite of crustacean limbs, though whether the animals fed in a similar fashion is unknown.

One pair of ambulatory appendages associated with the mouth have become modified to form mandibles by which the animal can bite, chew and manipulate its food, prior to swallowing it. The ability to move the limb in a plane transverse to the anterior-posterior body axis so that it can bite against its fellow in the mid-line with sufficient power and range of movement (gape) to allow feeding on large food particles has developed many times in the arthropods. The final products of these evolutionary lines are remarkably alike, but the similarity is due to convergence brought about by identical functional requirements in structures which were originally quite different.

In the Myriapoda and Insecta the distal podomeres of the limb were used to grip and handle the food. This action would be more efficiently carried out if the limb lost much of its flexibility and was strengthened by the disappearance of the articulations between the podomeres. In the scolopendromorph Chilopoda five articulated sclerotic rings can be distinguished in the mandibles; in the Diplopoda, three; in the Symphyla, two; and in the Pauropoda and Insecta, one. The mandible is formed from the whole limb, even though no sign of separate podomeres can be seen, and the biting surface represents the original distal tip of the limb.

In the Myriapoda the movement of the mandibles is derived from the depressor-extensor movements of the trunk limbs and from the ability of the distal segments to move away from, or towards, the mid-line. In the Insecta the prime movement of the primitive limb that has been used by the mandible (Fig. 4.6a) is the rotation about its long axis. This is used for

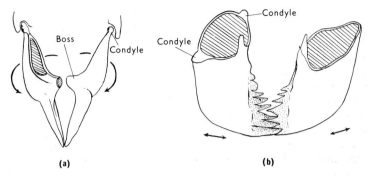

Fig. 4.6 (a) Mandible of *Machilis* (Insecta) articulated to head capsule by a single ball and socket joint; arrows indicate rotary movement which brings the mandibular bosses together and masticates the food between them. (b) Mandible of an insect articulated to the head by two ball and socket joints; arrows indicate the direction of movement, the animal being able to bite with the tips of its jaws.

rasping off the fine particles from its food. These may be further ground up by the action of medial projecting bosses of the mandible rolling together in the mid-line as part of this movement. This rolling movement utilizes the promotor-remotor swing of the limb, which becomes translated into a transverse movement by the lateral articulation of the limb moving posteriorly, and a second articulation developing anteriorly (Fig. 4.6b).

In the Crustacea the ambulatory limbs, by reason of the currents they created, already played a part in feeding. The accumulation of food particles in the mid-line and their forward transport to the mouth was aided by the development of a medial projecting boss, called the gnathobase, from the inner side of the coxopodite, present in the limbs of many Crustacea. Those around the oral opening are enlarged and play a special part in manipulating food into the mouth. Movements essential for this action make the limb less suited for locomotory purposes and the locomotory function of the limb becomes reduced and lost. The limb podomeres distal to the coxopodite atrophy and disappear or persist as a small mandibular palp (Fig. 4.3b). The mandible in Crustacea is thus formed from the basal podomere and not from the whole of the limb as in Myriapoda and Insecta.

Once the mandible is formed the same problems arise in its efficient use as occurred in the Insecta. Again the promotor-remotor swing is used and the food particles manipulated and crushed by the rolling of the gnathobases together. Again the posterior movement of the dorsal articulation and development of a second one serves to convert the movement from the anterior to the transverse plane.

A typical walking limb of a chelicerate (scorpion) (Fig. 4.3d), has seven podomeres called in sequence, coxa, trochanter, femur, patella, tibia, tarsus and pre-tarsus. The coxa arises from the ventral side of the body and has a large endite projecting towards the mouth. It has no point of articulation with the body and, except for the first pair, moves only slightly. The large extrinsic musculature of the coxa shows that, slight as the movement may be, it is strong and powerful. The trochanter is articulated by a pivot joint to the coxa and can move freely in any direction. The femur touches the trochanter anteriorly and posteriorly, permitting only a dorso-ventral movement. The femuro-patellar articulation is again dicondylic with horizontal axis; its main movement is downwards; it lacks extensor muscles. The patello-tibial joint is similar but has both extensor and de-pressor muscles. The tarsus is composed of a number of tarsomeres. The pre-tarsus forms a pair of claws and is similar to that found in insects but lacks the pulvilli and empodium. In the primitive Chelicerates the limbs play a part in manipulating the food in the region of the mouth as they do in Crustacea. However, the basic movement is quite different. In primitive arthropod limbs articulated by a single dorsal joint to the body wall, some small movement of the ventral side of the coxa into and away from the body is possible. This movement is repressed with the development of a second articulation. In Chelicerata however, the movement is exaggerated; the transverse manipulating movement of the coxal endites (gnathobases) exploits this factor and powerful muscles give it a strong transverse biting action.

Whatever their evolutionary origin, the actual biting lobes show common features imprinted upon them by the type of food they handle (Fig. 4.7). The cuticle covering the biting surface is thick and heavily sclerotized and frequently of a different colour from the rest of the lobe. In surfaces handling particulate food the biting surfaces tend to be flat and broad with some surface roughness to enable the food to be gripped. Strong heavy surface bosses occur where the food is crushed as with the gnathobases of *Limulus*, which can crush bivalve molluscs. In carnivorous arthropods the biting surfaces form sharp cutting blades or piercing points capable of penetrating into the prey. In herbivorous forms the mandible is blunt and heavy, capable of chewing and grinding the plant material. In highly evolved forms such as the wasps the mandibles may be used for manipulating the clay from which nests are built and end in spatular-shaped surfaces suit-able for this work.

The manipulation of food prior to its being swallowed is accomplished by several pairs of appendages. In primitive arthropods the gnathobases of the limbs surrounding the mouth contribute to this action, the limbs still retaining their locomotory functions. This situation persists in chelicerates, although in *Limulus* locomotion and chewing cannot both be performed at the same time. In mandibulate arthropods the appendages immediately following the jaws have lost their locomotory function and become specialized for food manipulation: these are the first and second maxillae. In detail their anatomy is complex and varied. Those found in insects must suffice as an example of this type of appendage.

The first maxillae of insects (Fig. 4.7) consist of a basic cardo, articulated

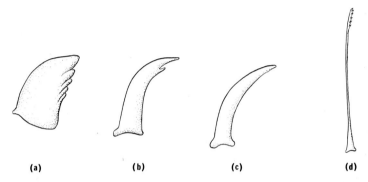

(a)   (b)   (c)   (d)

Fig. 4.7   Different types of mandible associated with different methods of feeding. (a) From a herbivore. (b) From a carnivore. (c) Piercing mandible with a groove on its inner surface. (d) Piercing stylet mandible from a mosquito.

by a ball and socket joint to the head capsule. Distally it forms a broad hinge joint with the stipes. Both these structures are sclerotized on their outer surfaces but have soft membranous cuticle medially. From the distal surface of the stipes arise the remaining podomeres which form a slender, usually five-jointed, maxillary palp of mainly sensory function. Medial to this, two lobes arise: an inner one, sharp-pointed and jaw-like,

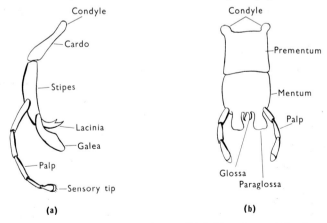

Fig. 4.8   (a) Maxilla of a biting insect. (b) Labium from a biting insect.

the lacinia; and covering it, a hood-like flexible broad lobe, the galea. The second maxillae (Fig. 4.8), labium of insects, is basically similar but is fused in the mid-line so that it acts as a lower lip to the buccal cavity. There is a basal plate, the pre-mentum, articulated by two ball and socket joints to the head capsule and movable only in an anterior-posterior plane. The stipes is similarly fused to form a broad distal plate, the mentum, also capable of movement only in the transverse plane. From its lateral distal edges arise a pair of three-jointed sensory labial palps. Two lobes on each side of the mid-line, the glossae, and lateral to these, a pair of paraglossae, are present.

In most arthropod groups the second maxillae do not fuse in the mid-line: a lower 'lip' to the mouth, if it exists, may be formed by fusion between other pairs of appendages. Still more posterior appendages may show adaptations to food gathering with or without a reduction in their locomotory function. These are the maxillipeds, their number and their depth of commitment to feeding activity differing greatly in the different arthropod groups.

Many arthropods use the more distal podomeres of their appendages for grasping and handling objects, mainly the capturing of prey and the manipulation of food. Some facility for grasping lies in the depressor extensor movements of the normal limb. In the praying mantis (*Mantis*), the depressor movement of the tibia against the femur is utilized in the modification of the first walking legs of this insect into a pair of powerful raptorial limbs. In these limbs the coxa is lengthened to give a long reach; the trochanter is small; the femur is swollen to accommodate large depressor muscles and it bears on its ventral surface two rows of spines; the tibia is more slender, bearing upon its ventral surface a single row of spines which, when the limb is folded, fits between the two rows on the femur. The tarsus is small but still functional since the limb retains some of its locomotory functions. The second maxilliped of the crustacean *Squilla mantis* (Fig. 4.9) is similarly modified, but here the extensor movement of the dactylopodite against the propodite is used, the grasping organ folding dorsally rather than ventrally. This difference is probably correlated with the burrowing habits of *Squilla* which strikes at its prey from underneath rather than from above as does the *Mantis*.

Where the grasping organ is formed so that the distal podomere folds back against the proximal one, the organ is termed a sub-chela (Fig. 4.9). These organs are widespread in the arthropods; they are common on the terminal limbs of Crustacea, especially Amphipoda, where they serve to pick up and convey food to the mouth, and in the Arachnida (Fig. 4.9). In spiders the first appendages, the chelicerae, are sub-chelate but the distal podite is pierced by a duct conveying poison nearly to its tip which is pointed and used to stab and inject poison into its prey as well as to grasp

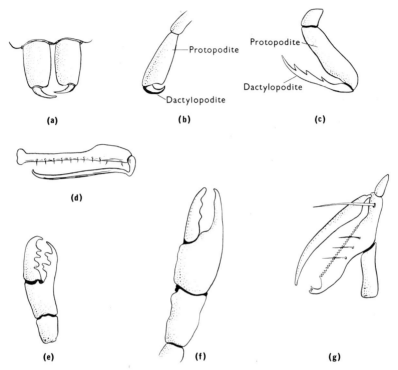

Fig. 4.9 Sub-chelate and chelate appendages. Sub-chelate: (a) chelicera of a spider, (b) thoracic limb of an amphipod, (c) maxilliped of the crustacean *Squilla*, (d) chelicera of the spider *Myrmarachne*. Chelate: (e) chelicera of a scorpion, (f) cheliped of a decapod crustacean, (g) anterior tarsus of a female wasp (Dryininae, Hymenoptera).

it. In the Myrmarachne the chelicerae are large delicate structures and are used by the male for carrying the female (Fig. 4.9d).

In chelae (Fig. 4.9), the grasping organ consists of a terminal podomere moving against a finger-like projection from the distal end of the next most proximal podomere. This arrangement greatly facilitates the capacity of the limb for the precise handling of the materials which it manipulates. It is widespread in the Crustacea and in the Arachnida but, like the sub-chelicerae, is of very infrequent occurrence amongst the Myriapoda and Insecta. In Crustacea the proximal podomere is greatly swollen to accommodate the muscles, both opening and closing, which operate the distal podomere. The term 'hand', sometimes used instead of chela, is very descriptive of its functions. In the Arachnida the structure is similar to that of Crustacea but the opening muscles are absent. The hand being opened by the elasticity of the cuticular hinge, powerful closing muscles are present.

As with the mandibles, the opposing surfaces of the podomeres of both the sub-chelicerae and chelae reflect their function. Heavy bosses are present where crushing and grinding work is done, sharp knife-like edges where cutting is the main function.

Many Crustacea spend most, if not all, their lives swimming, and their appendages are modified accordingly (Fig. 4.10). The limb has become flat

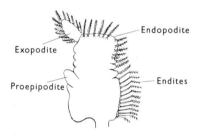

Fig. 4.10   Trunk limb, phyllopod of *Chirocephalus* (Crustacea).

and broad. The distinctions between the podomeres are obliterated, although a basic protopodite with endites and exites and a more distal endopodite and exopodite can still be distinguished. The swimming movement also creates water currents bringing food to the animal. The limbs are as much adapted to the collection and handling of food particles as they are for locomotion. The inner edge and distal extremity of the limb are fringed with long plumose bristles which are directed posteriorly. The limbs move in metachronal rhythm. On the forward movement water enters the space between successive limbs; on the back stroke the water is forced out of the space medially where the particles carried by it are filtered out and passed forward to the mouth. The backward movement imparted to the water propels the animal forward. Increased efficiency reduces the number of limbs necessary to collect food and the minimum size of particle that can be collected. *Daphnia* has but two pairs and can collect bacteria.

Appendages are also modified to subserve the reproductive activities of the animal. Two major types of modification occur:

1. Copulatory activities: in the male the appendages assume a cylindrical form and can be inserted into the female duct and internal fertilization can occur. The appendages may not belong to the segment on which the gonopore opens, e.g. dragonflies, spiders. Where paired reproductive

openings occur in the female each appendage in the male is separately modified: where one median duct occurs the male appendages form one half each of the copulatory organ.

2. In the female the eggs may remain with her until they are ready to hatch. The appendages are frequently modified so that

(i) they may be ensheathed with plumose hairs to which the egg mass will cling as in decapod Crustacea.

(ii) they may have a plate developed which, with similar plates from other limbs, forms a pouch against the ventral body wall, the eggs being retained in this space. The plates may appear in mature females and disappear when the animal is in a non-reproductive state.

(iii) the limbs may be used in courtship behaviour, chiefly in signalling between the sexes.

THE EXOSKELETON

The basic structure of the integument at the tissue level of construction has already been described; at a higher level of body organization it forms an exoskeleton, ensheathing the body in a system of stiff sclerotic plates joined by flexible membranes, and the appendages in cylindrical podomeres.

A sclerite comes into existence as a result of linkages being formed in the outer procuticle between protein molecules and between these and chitin. These produce an area of cuticle of medium rigidity, rather low tensile strength and moderate hardness. Weight for weight in these properties it compares quite favourably with a metal like aluminium. Variation in these properties to meet the mechanical requirements of an exoskeleton results from changes in the material from which a sclerite is made, and from the physical shape of the sclerites.

Two important variations arise in the materials from which the plate is made. Increased rigidity can and does result from the deposition of inorganic salts in the cuticle in addition to the normal tanning processes, e.g. the heavy exoskeletons of the large decapod Crustacea. At certain points in the exoskeleton, such as the wing hinge in insects or the setal base of some sense organs, near perfect elasticity is required and a special protein resilin is secreted.

Typically, the sclerotic plates which ensheath the body consist of a dorsal tergite and a ventral sternite on each segment joined laterally by a flexible pleural membrane. If the sclerites are of small size and/or the arthropod is aquatic, the intrinsic properties of the cuticle may be enough to retain the shape of the sclerite against the distorting forces of gravity, muscular pull, and internal growth pressure. Otherwise the plate is braced by a system of folds and ridges, or it is bent and curved. Few sclerites are perfectly flat, this being the form of a plate where its shape contributes least to its strength. Most are either concave-convex, a form which allows the inner area of the plate to be braced against its periphery, or bent, the line of the bend resisting distortion by forces acting at right angles to it. The larger the plate the greater its span and the more need there is for mechanical design to re-inforce the basic strength of its materials.

The plate may be strengthened by internal ridges (Fig. 4.11), an infolding of the epidermis whose surface form initially dictates the shape of the sclerite. The position of the ridge is often marked on the surface by a line

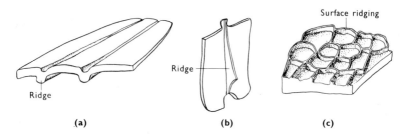

Fig. 4.11 Typical strengthening folds and sculpture of the skeletal plates. (**a**) Internal ridges as may occur on the thoracic tergite of wasps, (**b**) deep internal ridge from the pleuron of a pterygote insect. (**c**) Surface sculpturing.

called a suture. Internally the ridge serves for the attachment of muscles. More slender inwardly directed arms of the sclerite occur, the apodemes, for muscular attachment and are marked on the surface by a pit.

Internal folds may not always be desirable and strengthening of the sclerite may come from ridges and sculpturing on the external surface of the cuticle. These result from differential activity of the cells secreting the cuticle and not from folds of the epidermis. The surface patterns may form parallel ridges or mosaics of squares, oblongs, or hexagonal shapes, the sizes of which are sometimes sufficiently small to warrant the term microsculptures. Sometimes they are of sufficiently large dimensions to contribute obviously and materially to the strength of the plate (Fig. 4.11).

THE RESPIRATORY ORGANS

In many small arthropods, the integument is sufficiently permeable to gases to ensure an adequate supply of oxygen to the body tissues. With increase in size, with higher rates of activity, or with decreased permeability of the cuticle, this is inadequate and special organs develop which facilitate the entry of oxygen into the body. In aquatic arthropods these are gills; in terrestrial ones, lung books and/or trachea and tracheoles; in secondarily aquatic arthropods plastron respiration occurs.

The primary function of a gill is to provide a cuticular surface with minimal resistance to gaseous diffusion, of adequate area and placed in such a manner that the external fluid flows freely over its surface and the blood freely through its internal spaces. It is an advantage if external and internal fluids flow in opposite directions: then the internal fluid, when it leaves the gill, will be in equilibrium with the highest gaseous content of the external fluid.

The gill may be a limited area of cuticle, thinner than normal, with a well-developed blood flow across its inner surface. Such gills occur on the carapace of Crustacea, on the limbs of barnacles, and on the appendages of many other crustacean orders. Often the cuticular surface is extended by outgrowths, which may be simple leaf-like sacs with marginal blood vessels and much cross-vascularization, as in the amphipods and Anaspidacea. More frequently the gill has a central axis standing out as a triangular leaf from the body wall. From its surfaces there are other secondary outgrowths which may be lamellar, dendritic or filamentous (Fig. 4.12a). Contractile vessels may occur at the gill base to increase the blood flow through the organ. Control of the water flow across the gill comes from the geometry of the gills and surrounding integumental folds.

In *Limulus* the gills are borne on the posterior surface of posterior appendages; each consists of a large number (80 or more) of close-set lamellae; each lamella is a thin semi-lunar plate, the margin of which is rigid and fringed with setae, and is attached transversely at its base (Fig. 4.12b).

In the lung books of terrestrial Arachnida (Fig. 4.12c) the respiratory surface is also a number of close-set thin plates held apart by sclerotized projections from their surfaces. These become enclosed within a cavity by growth of the surrounding body wall out over the respiratory surface. Only a narrow opening, the spiracle, allows communication between the cavity of the lung book and the air. This spiracle may be opened and closed by special muscles and the cavity expanded or contracted in volume, thus pumping air in and out of the 'lung'.

The tracheoles are the sites at which oxygen enters the body in many terrestrial arthropods. They are present in what may be their most primitive condition in *Peripatus*. The external openings of the tracheoles lie very close to the body surface, a number opening into a shallow intucking of the body wall, the spiracle, which is just deep enough to penetrate the basement membrane (Fig. 4.12d). A large number of spiracles are present, 75 per segment, scattered over the body surface. No closing mechanism is present, and the loss of moisture from the ends of the spiracles causes the animal to lose up to a third of its weight when in atmospheres of less than 90% relative humidity (R.H.). Studies of the

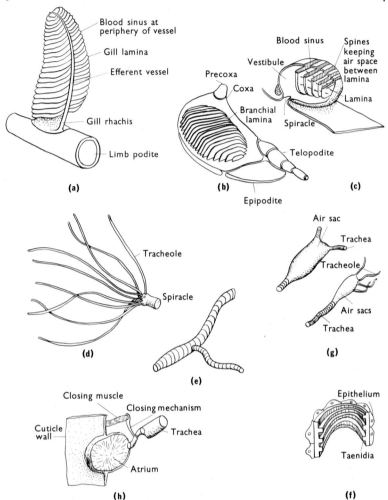

Fig. 4.12 Diagrams of the respiratory organs developed from the cuticle. (a) Gill of a decapod Crustacean. (b) Gill of *Limulus*. (c) Lung book of a spider. (d) Spiracle and tracheoles of *Peripatus*. (e) Part of a trachea. (f) Section showing structure of trachea. (g) Two types of air sacs developed from trachea. (h) Spiracular closing mechanism of a louse. (Redrawn from Harris, 1915)

evolution of arthropods indicate that the tracheal system must have arisen a number (perhaps 11) of times. While it is convenient to suppose that in each case its primitive form may have been like that shown in *Peripatus*, there is no evidence that this was in fact the case.

The further elaboration of tracheal respiration leads to a complex system in which the following structures may be considered as organs:

1. The trachea themselves. These are branching tubular ingrowths of the body wall. They have much the same structure as the integument. Internally the trachea have a spiral lining, the taenidia (Fig. 4.12f), consisting of thickenings of the endocuticle, each forming a few turns of the spiral. Taenidia help the tube to maintain its shape against pressures of the internal fluid and allow some facility for it to expand and contract longitudinally when pulled or pushed by the changing positions of the body organs which they supply. Little or no gaseous exchange takes place in a trachea.

2. The air sacs (Fig. 4.12g). Often the trachea is expanded in places to form a thin-walled sac.

3. Spiracular closure mechanisms (Fig. 4.12h). The opening into the tracheal system develops from the simple holes of *Peripatus* to become pits from which filtering setae keep particles out, and are equipped with a closure mechanism.

Where the animal has returned to aquatic life re-adaptation to aquatic respiration has followed two lines:

1. Tracheal gills. These have much the same general structure and function as ordinary gills but the oxygen, once through the cuticle, enters the tracheal system and passes to the body cells via these and not by the blood stream.

2. In association with a spiracle or covering the surface of a gill, a special area, the plastron, may develop. This consists of a surface covered with minute trabeculae or close-set hairs such that they present a hydrophobe surface so that water cannot penetrate between them. A thin film of air is trapped in this space. The absorption of oxygen from this layer decreases its tension compared with that in solution in the water. Oxygen from the water then passes into the air space and to the animal. Carbon dioxide in the plastron space can pass into the water six times more quickly than can the oxygen passing the other way. The air in the film requires to be renewed only occasionally by the animal's coming at intervals to the surface of the water.

THE INTEGUMENTAL GLANDS

These glands are widespread throughout the Arthropoda and are called tegumental glands or, in Insecta, dermal glands.

In its simplest form (Fig. 4.13) the gland is a single large cell with a thin protoplasmic filament penetrating the cuticle and opening to the external

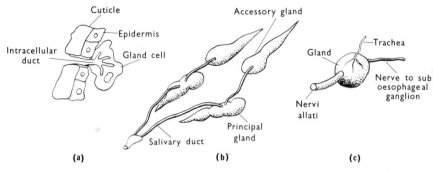

Fig. 4.13 Diagrams of the integumental glands. (a) Simple unicellular gland. (b) A more complex gland, salivary glands of *Notonecta* sp. (Hemiptera, Insecta). (c) Corpus allatum of an insect, an endocrine gland developed from the epidermis.

surface. An intracellular duct may develop within the protoplasmic filament. They may be very abundant, as many as 400 per mm² being present in some ticks. Typically, the gland consists of glandular and duct cells, the latter producing either an intracellular duct, or, by assuming a cylindrical shape and secreting a duct on its inner surface, an extracellular duct. Both types may be present, the intracellular duct being contained in cytoplasmic filaments of the duct cell which penetrate into the substance of the gland cell, and deliver the secretion into the extracellular duct which conveys it through the cuticle to the outside. An enveloping cell encloses the gland on its inner surface.

The products of these glands are very diverse: they secrete the outer protective layer of the cuticle, produce adhesive materials, attractant and repellant substances, poisons, and mucus-like substances.

With increased requirements for their products the glands get larger and more complicated (Fig. 4.13b). The number of glandular cells increases and the gland then projects into the haemocoele; the duct cells increase to form large branching tubes which may be swollen to form a sac-like reservoir. The gland may retain a simple tubular shape or be branched with small glandular pockets at the tip of each branch. The former design allows, in association with the appropriate muscles and nerves, the ejection of considerable quantities of material at any one moment: the latter favours continual copious secretion unless a reservoir is present when forcible ejection of large quantities also becomes possible.

Where the glands are greatly enlarged their number becomes reduced, often to a pair on a segment, as for example, the labial salivary glands of pterygote insects or the accessory glands of the male and female reproductive systems.

In a few cases the glandular cells have lost their connection with the epidermis and do not secrete directly to the exterior, such as the corpora allata (Fig. 4.13c) of insects. The corpora allata cells arise from the epidermis of the maxillary segment and move towards the gut behind the brain; here they associate with nerves and form endocrine organs.

THE EXERTILE VESICLES (Fig. 4.14)

These organs are widespread among the more primitive members of the different arthropod groups. They occur in a ventral position on abdominal segments of the Myriapoda, Insecta and Arachnida, but in the Onychophora

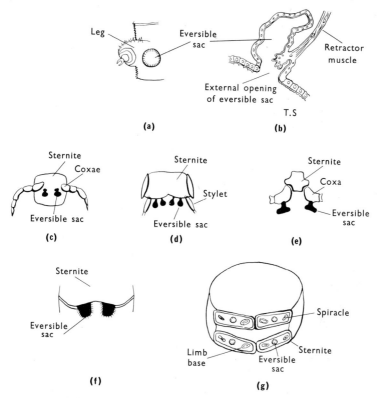

Fig. 4.14 Diagram of exertile vesicles from arthropods. (**a**) Extended vesicles from the base of a trunk limb of *Peripatus*. (**b**) Section of the above but with the vesicle retracted. (**c**) Extended vesicles on a trunk segment of *Scutigerella* (Myriapoda). (**d**) From an abdominal segment of Thysanura (Insecta). (**e**) From the coxa of a pair of trunk limbs in a millipede. (**f**) From the abdominal sternite of a pedipalp *Amblypygi* (Arachnida). (**g**) Reconstruction of a segment from the Carboniferous millipede *Acantherpestes*. (Redrawn from Fritsch, 1899)

they occur in the coxae and are the coxal glands. These organs are thin-walled sacs opening to the exterior through a narrow slit or pore. They can be everted by blood pressure when they form silvery bladders projecting from the body wall; they can be drawn back into the body by retractor muscles. They are organs for water absorption.

THE SENSE ORGANS

The primary sense cells are modified to respond to energy produced by molecular, atomic and sub-atomic forces, by generating a series of nerve impulses which they then transmit to the central nervous system. Thus photons are detected as light and heat, chemical diversity through molecular architecture and valency forces, movement by movement, which in its most sensitive manifestation is the movement of particles of molecular and atomic size. The primary sense cell also acts as an amplifier: the degree of amplification may be so great that in extreme cases a single photon may initiate a message. Primary sense cells are almost always associated with other cells to form sense organs. The function of these other cells is multiple; they protect the sense cell from undue stimulation, allow only certain types of stimulation to reach, and limit the amount of force that can be applied to it. By their structure and arrangement about the sense cell they build relationships between the basic stimuli and cell responses which are adaptive to the requirements both of the organism and of the environment.

The basic properties of the primary sense cell form a suitable basis for grouping sense organs into mechanoreceptors, chemoreceptors, photoreceptors, and temperature receptors although the actions of the associated cells may cause a response to environmental stimuli not directly related to these.

Mechanoreceptors

The simplest mechanoreceptors (Fig. 4.15a) are the movable setae which are widely scattered over the body surface and respond to any force such as touch, air movement or pressure change which can cause their displacement. Each consists of a single cell (the trichogen) produced distally into a fine cytoplasmic filament covered with cuticle where it projects from the body surface. The cell body and the non-cuticularized base of the filament are ensheathed by another cell, the tormogen, which isolates it from the surrounding epidermal cells and is covered with fine cuticle. The setae can be moved in any direction. This movement is detected by a primary sense cell whose distal process is inserted at one place at the base of the setae. A fourth cell encloses the proximal ends of the trichogen, tormogen and sense

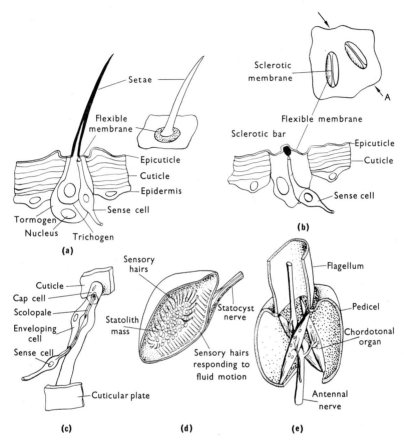

Fig. 4.15 Diagram of the sense organs of some arthropods. (a) Surface view of transverse section of a simple sensillum trichodeum. (b) Surface view and transverse section of a sensillum campaniformium. (c) Sketch of a chordotona sensillum. (d) Statocyst of a decapod crustacean. (e) Johnston's organ from the base of the antenna of an insect.

cell, separating them from the haemocoele. These organs are known as sensilla trichodea.

The thin flexible cuticle which surrounds the setae isolates the sense cell from being affected by movements of the surrounding exoskeleton and ensures that displacement of the setae comes only through external stimuli. Detection of movement of the exoskeleton by forces applied to it results from stimulation of another type of mechanoreceptor known as sensilla campaniformia (Fig. 4.15b). These organs are formed from a single cell lying in an oval slit in the exoskeleton. The surface of the cell domes outwards from its edges and is covered with thin cuticle. Compression, particularly in the line of the long axis of the oval, will cause the cuticle to bulge out, while stretching will cause it to be flattened. This movement is detected by a primary sense cell whose distal process is firmly attached to the centre of the cuticular bulge. The tip of the sense cell is usually much more elaborate than that of the sensilla trichodea. Added direction and precision can be given to the movement detected, by development of a thick bar over the organ cell in the long axis of the oval. Given different elastic properties of the exoskeleton, the organ cell cap and the rod form a very efficient amplifier, the tiny movements stress induces in the exoskeleton being transformed into considerable movement of the dome surface.

The chordotonal sensilla of Insecta (Fig. 4.15c) and the myochordotonal organs of Crustacea are internal organs whose function is the detection of movement of one skeletal plate relative to another, or of an internal organ relative to the exoskeleton. Each sensillum consists of three cells; one, a primary sense cell whose distal process comes in contact with two, a cap cell which is a modified epidermal cell, and is ensheathed by, three, an enveloping cell. Usually the cap cell is attached to one structure, the enveloping cell to another: movement of one structure relative to the other is transmitted to these cells and detected by the sense cell.

The sense cell may fire continuously, although the impulses are for most of the time infrequent and of low potential; or the cell is often 'silent', the burst of impulses being separated by long periods of quiet. In the former (postural) organs the firing may facilitate certain neural junctions and thus keep in readiness neural pathways within the central nervous system. In both cases a message is generated by a change of stimulus, the message commencing as the stimulus alters and either ceasing when the stimulus stops changing, or continuing to be produced for a period after the stimulus stops changing; it will then cease if the stimulus is removed. The period for which the message is generated has been observed to go on for as long as twenty minutes but it is frequently shorter than the period for which the stimulus is applied. In the case of the silent neurons the message can only be a burst of nerve impulses; for continuously firing neurons it can be a burst of greater frequency and potential or a reduction in number and strength. This gives a very useful sensing property for factors like temperature and pressure where both increases and decreases of a constantly present stimulus need detection. These properties of messages apply not only to mechanoreceptors but to other sense organs as well.

The simple mechanoreceptors described above form organs widely distributed throughout the arthropod body each capable of sensing events and feeding the necessary information to the central nervous system. Further elaboration occurs to give integrated action between the receptors by the following modifications:

1. The population of organs becomes very dense, individual sense organs having different properties so that a surface capable of fine discrimination and diverse responses is formed. Cuticular proprioceptive hair plates are of frequent occurrence where one sclerite overlaps another and the relative positions of the two need to be accurately signalled. Groups of campaniform sensilla on the coxa of the leg of *Periplaneta* or the chordotonal sensilla at the wing base of the fly *Empis* are examples of this grouping.

2. The population of neurons increases and may have diverse properties but their surrounding cells are themselves modified so that a greater range of stimuli can be detected and greater efficiency in analysis achieved. The principal organs in this group are the statocyst organs of Crustacea, Johnston's organ in the base of the antennae in Insecta, pressure receptors, and sound receptors.

The statocyst (Fig. 4.15d) is restricted to the higher Crustacea. It consists of an invagination of the body wall to form a spherical sac open to the environment at the site of invagination. The sac contains fluid and one or more inert bodies, often sand grains cemented together, the statolith or statoliths. The sensory field within the sac is formed from trichodea sensilla. Movement of the animal is not transmitted at once to the statolith or to the fluid within the statocyst. This inertial quality makes them, for a moment, a stationary point of reference against which the speed, duration and direction of the motion can be signalled to the central nervous system. In addition, the statolith is held in a definite position in the statocyst by gravity, so that changes in position of the animal relative to the gravitational field can be monitored.

Johnston's organ (Fig. 4.15c) is basically a cylinder formed of chordotonal sensilla lying in the pedicel of the antennae. Each sensillum is attached proximally to the base of the segment and distally to the membrane between the pedicel and the flagellum. Where small numbers of sensilla are present the organ forms a series of separate bundles set around the circumference of the pedicel. The organ is able to detect any movement made by the flagellum relative to the pedicel, including torsional movements, and to signal to the central nervous system about the strength and time course of antennal deflexion.

One form of motion of a particle is a repeated to-and-fro movement about an equilibrium position. In a continuous medium this vibration is communicated to neighbouring particles by elastic forces. From a point source this movement is transmitted through the medium in all directions at approximately equal rates, so that all particles on a spherical contour will be in the same state of vibrational movement. At any one contour over a period of time this will appear as a number of compression waves. Arthropods may detect either the extent of the vibration of the particles, that is the amplitude of the waves; or the force with which the particles are displaced, which is the pressure they exert on a surface. All sound receptors have the following properties: they must be bulky enough to be influenced, not by the random movements of individual particles, but only by the concerted action of a large number of them, and they must be able to oscillate with the wave movement.

The simplest organs capable of detecting compression waves are the trichodea sensilla which are light enough to be set in oscillatory motion by these forces. In Crustacea they appear to be the sole organs of sound detection, but in terrestrial arthropods they are supplemented by more complex organs, the tympanic organs, whose basic detector is the chordotonal sensillum.

The tympanic organs of insects are all designed to detect the displacement of particles. This is done most efficiently by a diaphragm whose mass is slight and whose articulations offer little resistance to movement. The maximum movement of such a diaphragm by sound waves occurs if the incidence of sound is at right angles to the plane of the moving element. (This differs greatly from the mammalian ear which is a pressure detector and as such has to be enclosed in a massive sound 'proof' chamber whose opening to the exterior is enclosed by a stiff almost immovable membrane. The orientation of pressure receptors with respect to the source of sound is unimportant.)

The tympanic membrane of the insect receptor is relatively large. It consists of a thin outer cuticle lined by a thin epidermis. The exoskeleton surrounding it is often heavy and well sclerotized, forming a thick frame holding the membrane. The inner side is formed from an air sac. The two become closely fused to form a thin light membrane which has both an inner and outer surface exposed to air, giving the least possible resistance to movement. Tympanic membranes are found on the abdominal segments of grasshoppers, cicadas, and many moths. Often they are further protected by sinking down into the haemocoele and by outgrowths of the cuticle extending over them. This both protects them from damage and restricts the field of sound entering the organ, helping accurate location of the sound source.

The membranes so formed are able to respond to a wide range of sound frequencies from as low as 300 to as high as 100 000 cycles per second, though for any one organ the range is less than this.

The chordotonal sensilla present are attached to the tympanic membrane by their cap cells and often, but not always, to some nearby rigid skeletal plate by their enveloping cells. Studies indicate that harmonic analysis of the sound, that is, discrimination of the different sound frequencies by different sensilla, does not occur. The sensilla do respond to changes in intensity of sound, and the sounds produced by insects differ most importantly in periodic changes of intensity (modulation). A sound needs to be heard for an appreciable period of time for this characteristic to be apparent. The organ does not readily fatigue, no diminution being observed for periods of stimulation of up to 30 seconds.

Chemoreceptors

The primary sense cells of these organs can discriminate between different chemical species and the magnitude of the response is proportional to the concentration of the applied chemical stimulus. The mechanism of discrimination is not known; it is clear however that different sensilla have different discriminatory powers. For example, in one chemoreceptor from the blow-fly, one sensillum responds to carbohydrates, another to monovalent salts. A chemoreceptor organ rarely has only one sensillum: it usually has two, three or four, and often many more are present. A single epidermal cell is modified to produce a special area of cuticle through which the tips of the sensilla are exposed through tiny pores to the external environment (Fig. 4.16). The cell bodies of the sensilla are grouped to-

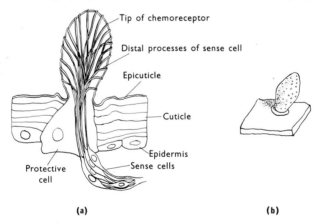

Fig. **4.16** Chemoreceptors. (a) Transverse section through a typical chemoreceptor sensillum basiconicum. (b) Surface view of same.

gether close to the base and their long narrow filaments form a bundle penetrating the cuticular cell: the filaments may separate from the bundle and pass to the cuticular pores which are separately spaced, or they may continue as a bundle to open to a common pore. Another cell lies over the base of the organ and separates it from the haemocoele.

The actual shape of the cuticle produced by the epidermal cell and the arrangement of pores on its surface is very variable. Maximum exposure to the environment is achieved by peg-like projections perforated with numerous pores; minimal exposure by restriction of the pores either to a single one at the tip or to a row around the circumference at its base, or by the organ being sunk into a pit which has a restricted opening.

Sensitivity may be very great, a few tens of molecules impinging upon the organ being sufficient to stimulate it. This extreme sensitivity is found in terrestrial arthropods where very low concentrations of molecules in air are to be detected. These olfactory organs can be contrasted with chemoreceptors where contact with material dissolved in water is essential and which are less sensitive. These are the chemoreceptors of aquatic arthropods and occur also in terrestrial arthropods where material in solution is to be sensed. They are referred to as contact receptor organs rather than taste organs since they are not necessarily associated with the mouth.

The eyes of arthropods

The basic light receptor cell in arthropods is similar to that found in other animals. It is in the array of these cells to form more complex organs that the eyes of arthropods differ so greatly from the eyes of other animals.

The visual organs of arthropods are in detail fairly diverse but they can be grouped into the ocelli, the ommatidia, and the compound eyes.

The ocelli

The light receptor cells of the ocelli are arranged in a shallow cup, the light receptor parts of the cells set perpendicular to the cup surface and facing towards its centre, the cell bodies and axons orientated away from the light source (Fig. 4.17). In a few ocelli the cell body lies between the distal process and the light source, an inverted retina being formed as in vertebrate eyes. In some arachnids both types of retina may occur in a single ocellus. The epidermis lies between the retina and the transparent cuticle which covers the eye. Its cells are well formed and are more transparent than the surrounding cell sheet. The cuticle may simply be a transparent area covering the eye, but is often thickened to form a massive biconvex lens. A basement membrane separates the retina from the haemocoele.

The ommatidia (Fig. 4.17c and d)

Each ommatidium consists of a number of retinula cells in which the sensory area is confined to the edge of the cell facing in towards a centre axis. Often this axis is filled with fluid, the light-sensitive faces of the cells

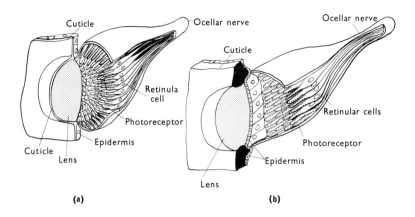

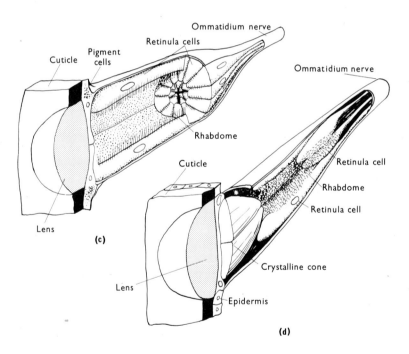

Fig. 4.17 Diagram of the structure of the visual organs of arthropods. (**a**) Section through the eye of *Peripatus*. (**b**) Section through the lateral ocellus of a moth *Zygaena*. (**c**) Section through the ommatidium of centipede *Lithobius*. (**d**) Section through the ommatidium of an insect *Lepisma*.

then forming a number of separate vertical columns. In many ommatidia the cells touch each other in the mid-line, the light-sensitive faces then forming a column called the rhabdome. Normally six or seven cells enter into the composition of the rhabdome: usually however one cell does not, nor does it bear a light-sensitive face; this is the excentric cell. Proximally the retinula cells are separated from the haemocoele by a basement membrane through which their axons pass separately to the optic lobes of the brain. Laterally they are ensheathed by pigment cells in which the pigment is capable of being moved either to enclose the retinula complex completely or to contract to one end and expose them to light entering from the side.

Distal to the retinula cells lies the crystalline cone, a conical-shaped transparent body formed from four cells. The tip of the cone is in close association with the tip of the rhabdome, the broad base orientated distally. The crystalline cone almost touches the cuticular lens, a hexagonal-shaped biconvex transparent region of the cuticle. Separating the lens from the crystalline cone is a thin film of cytoplasm belonging to the epidermal cells which secreted the lens. The cell bodies form a sheath enclosing the crystalline cone. This part of the epidermal cell contains pigment which is also capable of being moved to be dispersed in bright, and contracted in dim, light.

Compound Eyes

Each compound eye consists of a number of ommatidia associated together to form a complex organ (Fig. 4.18). This association has occurred independently in different evolutionary lines. In the more primitive animals eyes have large lenses and shallow photoreceptors. In advanced forms the lenses are reduced in size, have greater uniformity in diameter and the photoreceptors are deep. The centipedes *Lithobius* and *Scutigera* and the apterygote insects *Lepisma* and *Petrobius* show the differences between these conditions. In the holocroal eye the lenses of the ommatidia touch, forming a continuous transparent surface covering the organ. In the schizocroal eye the lenses are separated by areas of opaque cuticle. Both types of eye are found in the Trilobita.

The surface of the compound eye (Fig. 4.19) is always convex, often slightly so, but frequently it forms a hemisphere projecting from the surface of the body. The optical axes of the ommatidia are set obliquely to one another so that each ommatidium is 'looking' at a different part of the visual field. Considerable overlap of the fields of vision of adjacent ommatidia occurs, an overlap which is greater the greater the convexity of each individual lens.

The eye is capable of detecting pattern, and forms an image of the animal's surroundings which falls on a receptor surface. The manner in

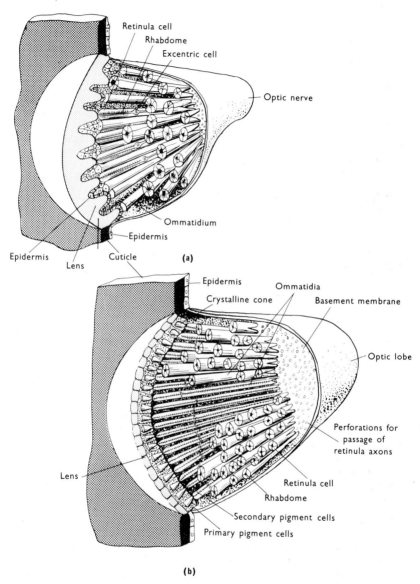

Retinula cell
Rhabdome
Excentric cell
Optic nerve
Ommatidium
Epidermis
Epidermis
Lens   Cuticle   **(a)**

Epidermis
Crystalline cone
Ommatidia
Basement membrane
Optic lobe
Perforations for passage of retinula axons
Lens
Retinula cell
Rhabdome
Secondary pigment cells
Primary pigment cells

(b)

Fig. 4.18 Compound eyes in arthropods. (**a**) Schizocroal eye of *Limulus*, showing widely spaced ommatidia with well marked excentric cells, crystalline cones are absent, and the lens is formed by proximal cone-like projections of the cuticle. A not dissimilar eye occurs in the glow-worm (Insecta, Lampyris). (**b**) Holocroal eye typical of most Insecta and Crustacea. The ommatidia closely packed, excentric cells if present not well marked, crystalline cones present often hard occasionally soft and jelly like, the lens formed from the whole cuticle.

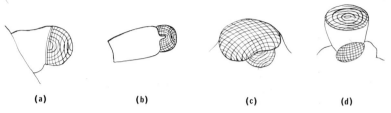

(a)   (b)   (c)   (d)

Fig. 4.19 Diagrams of surface of some compound eyes. (a) Eye of a zygopteran dragonfly. (b) Eye of a decapod crustacean. (c) Split eye of a fly (*Bibio*). (d) Split eye of a mayfly (*Baetis*). In (c) and (d) the lens sizes of the ommatidia of the two parts of the eye are of different diameters.

which the compound eye accomplishes this is unique in the animal kingdom.

In eyes in which each ommatidium is optically isolated from its fellows, e.g. a holocroal eye in which heavy pigment is fully dispersed, the visual capacity is largely the sum of the properties of the individual ommatidia. Although the effective image for each ommatidium is formed at the tip of the crystalline cone, it is doubtful if the details of the image can be resolved by the rhabdome, the stimulus being simply a spot of light. The intensity of this spot will be different in the different ommatidia according to variations in illumination of the different parts of the visual field which they face. A dark area in the visual field will give an area of ommatidia less intensely illuminated than its surroundings. In this manner a mosaic of spots of light of different intensity will be formed at the distal ends of the rhabdome, the light spots forming a crude image of the environment. The resolution of the image will be greater the larger the number of spots formed, hence it is greater in eyes in which the most ommatidia 'look' at a given area, i.e. the flatter is the eye and the more nearly parallel are the optical axes of the ommatidia.

In holocroal eyes in which the ommatidia are not optically isolated from each other, e.g. the pigment is retracted or the light so intense that it penetrates the pigment sheath, the optical properties of the lens and cones of adjacent ommatidia permit the formation of refraction images deep in the eye. A succession of images lying further and further away from the lenses occurs but only the second and third, which lie approximately half-way along and close to the proximal region of the photoreceptors, are effective. These second and third images confer a higher resolving power on the eye than does the first, mosaic, one.

All arthropod eyes lack a focusing mechanism; it is possible that for nearer objects the depth of the rhabdome compensates to some extent for this, the images varying in their precise position on the rhabdome with different distances of the object from the lens.

THE TUBULAR ORGANS

In these organs the basic plan is that of an epithelial cylinder through which material flows. Although of diverse origin and function, organs built to this design have many features in common. Two main types occur: the organs of sequential processing (the gut and coelomic organs), where material as it passes through them is subjected to an ordered and well-integrated series of processes; and the distributive organs (the heart, blood vessels and trachea) where the tubular design serves simply to conduct material unchanged throughout the body.

The organs of sequential processing

In these organs there is a site of entry of raw material, a uni-directional flow over the sequence process sites where addition or subtraction to the main mass can occur, and a site of exit where the processed product or the waste material leaves. Some of the processing sites may form diventricula, or tubes and pockets, adding to the complexity of the organ.

There is a mechanism for moving material through the organ. In arthropods this is almost always muscular, either by peristalsis of an ensheathing muscular coat, or pressure generated by contraction at one end of the organ. In a few cases movement may be generated by pressure resulting from the continuous secretion of material at a given point.

The flow of material is not constant throughout its length but varies, the rate being adapted to the type of process occurring there. This flow rate is regulated by valves: one almost always occurs at the exit site, others elsewhere in the tube, frequently where there is a major change in flow rate.

Certain regions of the epithelial tube are enlarged to form storage sites. These serve to hold materials accumulating by the fast processing sections of the organ prior to passage through the valve to the slower regions, or to store materials where the behaviour characteristics of the animal dictate intermittent entry or exit of materials from the organ. Storage sites are not necessarily 'idle'; mixtures with materials from active sites result in changes occurring in them.

The alimentary canal

The function of the alimentary canal is to process the material entering the mouth, extracting from it the nutritious material, absorbing this and passing out of the body the indigestible portion, together with the waste products of the body's metabolism.

The gut of *Peripatus* (Fig. 4.20a) is one of the simplest types found in the Arthropoda. The food, consisting of the tissues and juices of small invertebrates, is rasped from the body of the prey, fragmented by the jaws

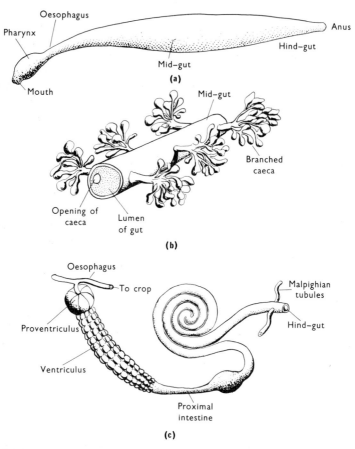

Oesophagus

Pharynx

Anus

Hind-gut

Mid-gut

(a)

Mouth

Mid-gut

Branched
caeca

Opening of
caeca

Lumen
of gut

(b)

Oesophagus

To crop

Malpighian
tubules

Proventriculus

Hind-gut

Ventriculus

Proximal
intestine

(c)

Fig. 4.20 (a) General structure of a simple arthropod gut (*Peripatus*). (b) Mid-gut of an arachnid showing the large tubular lumen that can act as a reservoir with lateral branching tubules and terminal caeca where digestion and absorption occur. (c) Complex mid-gut showing external differentiation (Calliphora, Diptera, Insecta).

and mixed with saliva before being passed into the mouth. The food then passes back through a simple cuticular lined tube, the fore-gut, formed by an ingrowth of the body wall. Thick muscular bands around it just behind the mouth form the pharynx, which acts as a valve and as a pump forcing the semi-liquid food posteriorly. The posterior flow of the food is helped by peristaltic action of the muscular coat. The fore-gut is short, the alimentary canal being continued as the mid-gut which is of endodermal origin.

The posterior part of the fore-gut projects into the anterior lumen of the mid-gut and is folded back on itself, the mid-gut arising from the anterior end of this collar. This forms a simple valve regulating the passage of material between fore- and mid-gut. The mid-gut is a broad cylinder of epithelial cells secreting on its inner surface a tubular peritrophic membrane. This encloses the food, separating it from the epithelial surface which it protects from abrasion by the food particles. Within the mid-gut the food is digested to give simple carbohydrates, amino acids and fatty acids which are absorbed, the indigestible waste remaining in the peritrophic membrane. The mid-gut cells produce and secrete the digestive enzymes, absorb the products of digestion and store the excess as glycogen, protein and neutral fat. Enough material can be stored to last a starving animal for nine months. The cells secrete the metabolic nitrogenous waste of the animal in the form of uric acid which is deposited upon the surface of the peritrophic membrane.

Posteriorly the mid-gut joins a short cuticular lined tubular invagination of the ectoderm, the hind-gut. The hind-gut opens at the posterior end of the animal through an anus which is closed by a pair of fleshy lip-like valves.

The peritrophic membrane with its waste food and uric acid crystals is ejected from the gut once every 24 hours by the animal's swallowing air. A fresh membrane is then secreted.

In other arthropods further anatomical elaboration occurs associated with the more efficient processing and changes in dietary and water content of the food. Considerable differences may be found, even amongst closely related species. Descriptions of a few types of gut are inadequate to outline the range to be found. Instead, an account of the functional principles underlying the anatomical modifications shown by the gut will be given. From these, charts can be prepared which cover a wide range of structure and function.

Modifications occur related to the size of the food particles passing down the gut. When they enter the mid-gut small size is an advantage allowing maximum speed in chemical processing. Food taken in at the mouth may consist of particles of the correct size but, as the mouthparts adapt to masticate larger food masses, they lose the ability to shred the food finely enough and to convey fine particles into the mouth. The food entering the mouth is then composed of particles exceeding the size that is required for optimal mid-gut action. In these conditions subsidiary grinding mechanisms are developed in the fore-gut. In the higher Crustacea the anterior part of the fore-gut is enlarged forming a nearly spherical sac lined with thick cuticle and having a powerful muscular coat, the gastric mill. A number of thick calcareous plates are present which can be moved by muscles and operate to grind the food into small particles. An elaborate filter chamber opens posteriorly from the gastric mill. This directs the soft finely ground nutritious particles into the mid-gut, and the coarser harder particles to a channel which conveys them through the short mid-gut and deposits them directly into the hind-gut. In Insecta the posterior end of the fore-gut just before its invagination in the mid-gut lumen forms a grinding chamber, the gizzard. This is surrounded by thick coats of longitudinal and circular muscle. The cuticle is thickened to form six longitudinal ridges bearing strongly sclerotized teeth anteriorly and setae posteriorly. The ridges can meet in the mid-line and occlude the gut lumen. In addition to their grinding action they serve as a valve regulating the entry of food into the mid-gut.

Many arthropods are fluid feeders and, in these, modifications occur in the fore-gut for sucking the fluid in at the mouth and pumping it to the mid-gut. In the Arachnida, which are all fluid feeders, the posterior region of the fore-gut is modified to form a pump, the sucking stomach. The lumen is enlarged and the muscular coat produces the pumping action. In the Insecta the pharynx is elaborated to form a pumping organ.

Changes take place in the anatomy of the mid-gut which facilitate the speed with which food can be digested and absorbed. In the mid-gut of primitive arthropods all the epithelial cells seem to have the same function and each individual cell is capable of producing and secreting enzymes, absorbing the products of digestion, reconstituting them in protein, carbohydrates and fats, storing these and excreting the nitrogenous waste of the body. A long mid-gut, periodically filled and emptied, is well suited to this kind of structure and appears widespread amongst the less specialized arthropods. The optimal diameter of the mid-gut is determined by a complex of factors, but two of the important features are:

1. The ratio of the surface area of the epithelium to the volume of the gut contents.
2. The relationship of the radius of the gut to another ratio, the ratio of (the time taken for the enzymes to diffuse from the cell wall to the centre) to (the length of time the contents remain in the gut).

In a cylinder the greater the bulk the less the surface area per unit weight of contents. Increases in the bulk of material that can be accommodated by the mid-gut are modified to try to keep this ratio as low as possible, either by increasing the length of the mid-gut or by the development of a much-pocketed epithelium or a few tubular caecae. These outgrowths usually have the same functions as the rest of the mid-gut epithelium. In the Arachnida (Fig. 4.20b) the caecae branch profusely and the alveoli at the tip of the branches are where the major part of digestion and absorption occurs. The tubular part of the mid-gut is relatively long and the caecal openings are widely spaced down it. In the higher Crustacea the mid-gut is very short, merely accommodating the openings of the caecae. This form of mid-gut not only retains the minimal surface/volume relationship but increases it, so that faster processing occurs.

In many arthropods the epithelial cells of the mid-gut do not retain all their primitive functions; correlated with this, those of similar function tend to occur together, thus forming functional sub-divisions of the gut. In the larvae of the mosquito *Aedes* sp. following a meal, fat droplets occur in the cells of the anterior end of the gut and glycogen granules in those at the posterior end. Storage of material in the cells still occurs in the guts of the Arachnida and Crustacea, but in the Insecta this is taken over by another organ, the fat body.

The aggregation of cells of specific function affects the cells and the anatomy of the gut. The advantages of the sequential processing this permits can only be fully exploited if the material is quickly passed from one section to another. High flow rate and efficient processing is accomplished by narrow long tubular mid-guts, the small diameter giving high surface volume ratio and the long length permitting fast movement while keeping the material in a specialized section for an appropriate duration of time. Mid-guts then, with highly developed sequential properties and good handling capacity, tend to be long and thin. Such guts are found in the blow-fly (Fig. 4.20c). They differ from the high capacity primitive mid-gut which tends to be long and thick and of uniform appearance.

The hind-gut conveys waste products to the anus where they are discharged to the exterior. In some specialized animals, termites for example, the hind-gut is enlarged to form a sac which contains trichonymphid Protozoa. These digest the cellulose of the animal's diet and the products of this are absorbed through the hind-gut into the animal.

Continuous feeding is not known to occur in any arthropod. Feeding at a rate equal to that at which the food can be processed does occur, but usually, due to intermittent feeding and unequal rates of processing, parts of the arthropod gut have a well developed storage function. In Insecta the fore-gut behind the oesophagus and in front of the gizzard is sac-like and can store a considerable quantity of food which is then passed on in smaller quantities over a long period of time to the mid-gut. In Arachnida the mid-gut is a long broad tube and considerable quantities of liquid food can be stored in it, enabling the animal to go for long periods without feeding. In Crustacea storage within the lumen of the gut seems uncommon, although the large volume of the ducts in much-branched mid-gut caecae may hold a considerable quantity of food. Storage of waste matter may occur in the hind-gut. In many insect larvae where defecation would foul the limited amount of food available, such as the bee larvae in cells containing pollen, or endo-parasitic forms, the hind-gut is closed and waste matter accumulates in it until growth and feeding are completed, then when the animal quits the site the contents are voided.

In many animals the gut is the main organ of nitrogenous excretion. However, in Crustacea nitrogen is excreted as ammonia and leaves the body via the gills. In the Arachnida where nitrogen is excreted mainly as allantoin and in the Insecta as uric acid, excretion is almost entirely through the gut. In Insecta long thin tubules, the Malpighian tubules, grow out from the junction of the hind- and mid-gut. The tubules are long and of small diameter and they penetrate throughout the haemocoele, ensuring that no tissue is very far from them. There is a light musculature, frequently a single fibre spiralling around them. In the living animal they are in constant motion. Sequence processing occurs along their length, abundant fluid containing ions and salts, especially potassium hydrogen urate, entering their distal regions, and fluids salts and ions being preferentially absorbed in their proximal region where the urates are deposited as uric acid. In Arachnida similar tubules also develop from the mid-gut. Valvular mechanisms guard the openings of the tubules into the gut.

The Myriapoda, Insecta and Arachnida are basically terrestrial animals. Hence the conservation of water is important to them. The terminal regions of the gut are modified for this purpose. The hind-gut absorbs water from the food and passes it into the haemocoele. In the Insecta the posterior part of the hind-gut (Fig. 4.21) has six pads of specialized epithelium developed around its lumen. Each pad consists of deep bi-nucleate cells set on a basement membrane; distally the cuticle they secrete is separated from their surface by a space which probably contains fluid in the living animal. At the periphery of each pad the cuticle is fastened to the surface of the cells. Between each pad the epithelium is composed of mono-nucleate cells, the

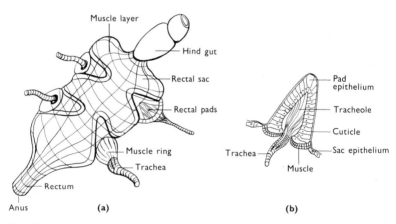

Fig. 4.21 Diagram of the rectal sac of an insect (*Calliphora*). (**a**) General appearance, the rectal pads are visible through the transparent wall of the sac. (**b**) Transverse section of a rectal pad (after Lowne, 1890).

cuticle firmly adherent to their surfaces. In the blow-fly *Calliphora* very large intercellular sinuses occur between the pad cells which communicate with the haemocoele through valve-like structures.

The Coelomic Organs

The coelomic organs (Fig. 4.22) of which, in primitive arthropods, there is a pair to each segment, consist basically of a thin-walled end sac opening through a ciliated funnel into a thick-walled convoluted tubule, which opens to the exterior through a pore in the ventro-lateral body wall. Both sac and duct are derived from mesoderm and represent remnants of the coelomic cavity and coelomoduct.

The organs on the different segments have different functions in *Peripatus*. The anterior pair are those of the third body segment and are the salivary glands. There is a small end sac and the coelomoduct is enlarged and straightened so that it extends back through several body segments. It opens into the buccal cavity. During feeding its secretion flows onto the food and some digestion occurs. The enzymes amylase, glycogenase and a protease are present. The saliva is swallowed back into the mouth.

The segmental organs in the trunk segments maintain the ionic balance of the haemolymph. The end sac has a thin-walled epithelium opening through a narrow ciliated tube into a convoluted tubule lined with thick columnar epithelium. This opens to a large bladder which opens through a narrow duct medially near the base of each leg. The haemolymph is filtered through the walls of the end sac, the motive power being the hydrostatic pressure within the haemocoele. The fluid then passes down the convoluted tubule. The urine in the end sac is similar to the haemolymph but lacks the protein and the amino acids of the blood. In the convoluted tubule most of the ions are absorbed: others are lost to the body. Discharge of urine takes place from all the segmental organs simultaneously. Discharges are not frequent; a second discharge rarely occurs within 24 hours. Analysis of the urine shows the presence of chloride, phosphate and sulphate with small amounts of ammonia. In 7 out of 9 cases the urine was hypotonic to the blood. The ammonia is derived from urea and aids in the secretion of sulphate and phosphate while conserving sodium and potassium.

Part of the coelomic organs of all segments and those of the posterior segment have become associated with the germ cells and are modified to form the reproductive organs of the animal. Those from one segment form the ducts, that from the immediate posterior segment the accessory glands.

This general pattern of coelomic organ specialization is maintained throughout the Arthropoda, the anterior ones having a salivary function, the middle ones an ionic and excretory function, and the posterior one a

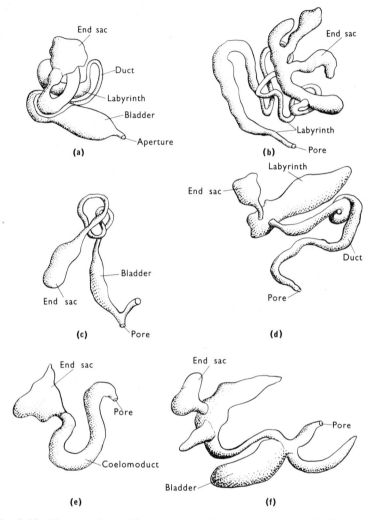

Fig. 4.22 Diagram of the different forms of the coelomic organs. (a) *Peripatus*. (b) *Limulus*. (c) *Machilis*. (d) *Galeodes* (Arachnida). (e) *Estheria* (Crustacea). (f) *Macropodia* (decapod Crustacea).

reproductive function. In all except Onycophora only one or two pairs of segmental organs remain.

In larval Crustacea the most anterior coelomic organ occurs on the third body segment: it is excretory in function and consists in the Branchiopoda of an end sac opening through a ciliated funnel into a short straight thin-walled tube with an exit pore on the base of the antennule.

In adult decapod Crustacea *Homarus* the antennary gland consists of a thin-walled sac formed by a single epithelial layer. The sac opens into a swollen tubule whose infoldings form a sponge-like labyrinth; at its distal end the labyrinth opens into a bladder which discharges through a narrow duct and pore provided with a closure mechanism on the base of the antenna. In freshwater crayfish a long tube occurs between the labyrinth and the bladder: in the crab *Macropodia* (Fig. 4.22f) the bladder is large and lobed and may penetrate between the other organs of the body. A blood vessel conducts blood from the heart to the segmental organ.

Fluid enters the end sac under hydrostatic pressure and is modified in its passage to the bladder. The urine is isotonic with the blood but contains higher concentrations of magnesium and sulphate and lower ones of potassium and calcium. Little nitrogen appears in the urine. Except in the crayfish the organ appears not to play a major role in osmo-regulation. The rate of flow of urine is from o·1 to o·6% of the body weight per hour in the lobster.

The coelomic organs of the sixth body segment in the Arachnida form the coxal glands, whose structure and presumably function are typical. In Crustacea they form the segmental organs in adult lower Crustacea, and in larval higher Crustacea. In insects they have a salivary function in *Machilis* but are absent from pterygote insects.

The gonads (Fig. 4.23)

In *Peripatus* the dorsal divisions of the posterior trunk somites fuse on either side to form longitudinal coelomic tubes which become the paired ovarian or testicular sacs in the ventral walls of which the gonads develop. Posteriorly to the paired gonoducts the last pair of somites meet at the mid-ventral pore.

In Arachnida the genital ducts develop as coelomoducts in the ninth segment. The somites of this genital segment unite dorsally and, with contributions from other segments, form the median gonadial sac.

In Chilopoda and Insecta the trunk mesoderm becomes sub-divided and consecutive dorsal lobes from a varying number of segments fuse to form two longitudinal tubes below the pericardial floor. Genital cells, which were previously segregated, sooner or later appear in the walls of coelomic sacs.

In Crustacea the gonads are formed in the same way but the coelomic spaces are small or obliterated early so that germ cells come to lie in a solid epithelium which later develops spaces not strictly continuous with the early coelomic cavity.

In those coelomic organs which contain the germ cells, the germinal epithelium becomes localized upon one wall of the end sac. Its products,

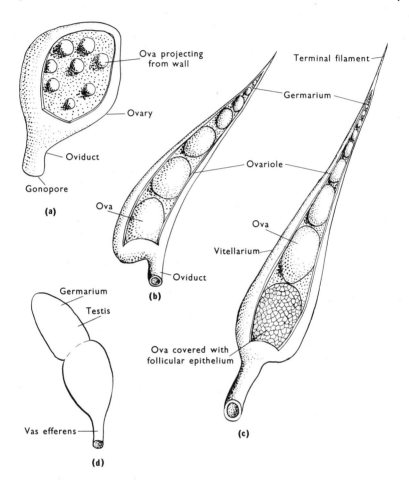

Fig. 4.23 The gonads of arthropods. (**a**) Simple sac like ovary with part of the wall removed showing ova developing from the wall of the sac. (**b**) Tubular ovary with part of the wall removed showing ova developing in sequence down the length of the tube. (**c**) Ovariole of an insect with part of the outer wall removed showing the ova covered with follicular epithelium at the distal end and germarium at the proximal end where the maturation divisions occur. (**d**) Testis.

the mature germ cells, are shed into the cavity and pass via the ducts to the exterior.

With the further elaboration and development of the oocyte before it leaves the organ the reserve food content of the egg is increased so that the hatchling is larger and better developed than would otherwise be the case. This is accomplished as follows:

1. The daughter germ cells associate together and their cytoplasm fuses: all but one of the nuclei disintegrate and a large oocyte is formed.
2. The follicle cells specialize for the passage of materials from the blood to the oocyte.
3. The oocyte remains adhered to the wall of the end sac which specializes for its continued nutrition.

The Circulatory Organs

The heart

The heart (Fig. 4.24) is basically a long muscular tube with a pair of lateral openings (ostia) and a pair of arteries in each segment. Blood enters

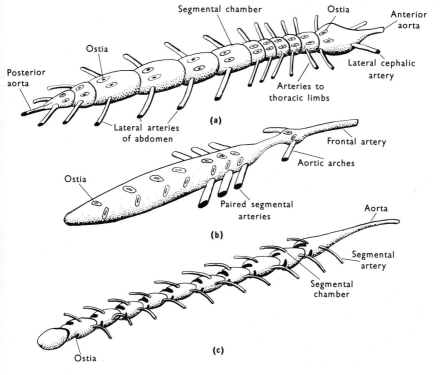

Fig. 4.24 Diagrams showing the hearts of arthropods. (**a**) *Squilla* (Crustacea). (**b**) *Limulus* (Arachnida). (**c**) *Periplaneta* (Insecta).

the heart through the ostia and leaves it through the arteries. Anteriorly there is a non-contractile tube, the aorta, which conveys blood to the head. Blood usually flows anteriorly but occasionally flow reversal occurs.

The muscular wall of the heart, the myocardium, is primitively composed of a sheet of muscle fibres, their long axes perpendicular to the anterio-posterior axis of the heart. In many arthropods the myocardium is composed of many layers of muscle fibres which may anastomose. The lining of the heart is formed from the sarcolemma of the muscle fibres; occasionally haemocytes may form a lining which resembles an endothelium, but a true epithelial lining to the heart appears to be lacking. Externally the myocardium is surrounded by a sheath of connective tissue.

The heart is held in position by elastic ligaments and muscles suspending it from the dorsal musculature and ventrally by a pericardial membrane attached dorso-laterally to the tergites. The part of the haemocoele containing the heart is the pericardial cavity. Muscles occur in the pericardial membrane; these arise from a point attachment on each side of the tergum and radiate out to a broad insertion beneath the heart.

The function of the heart is to provide motive power for the circulation of the blood. This is accomplished in two different ways.

In the tubular heart of *Limulus* and in crustacean hearts, systole, contraction of the heart, occurs as a single beat. Diastole, filling of the heart, follows each systole, expansion of the heart being accomplished by relaxation of the myocardial muscles and the pull of the elastic suspensory ligaments. These movements are correlated with the opening and closing of the ostia and arterial valves. At the commencement of systole the ostia close and the pressure of blood within the heart rises, the cardio-arterial valves open and blood flows out of the heart through the arteries. At the end of systole the ostia open and the arterial-cardiac valves close; diastole then results in blood entering the heart from the haemocoele. The contraction of the myocardium and its integration with movement of the valves depends upon nervous stimulation.

In the tubular hearts of insects a wave of contraction passes forward along the heart from the posterior end. Blood is carried along in front of the wave, most leaving the heart through the aorta to discharge into the haemocoele in front of the brain. Diastole, filling of the heart behind the peristaltic wave, comes from expansion of the heart by the elastic suspensory ligaments, assisted by the alary muscles in some cases, but the general function of these muscles is to move the heart within the pericardial cavity. The ostia close in front of the wave, thus preventing blood escaping laterally, and open after the wave, allowing the entry of blood from the haemocoele. In some insects the generation of the contraction wave also depends upon nerve stimulation: in others the heart will continue to beat after the nerves to it have been cut.

The blood vessels

The definite blood vessels of arthropods are arteries conveying blood from the heart to the tissues. Return veinous flow is through the haemocoele. In the large arteries there is a cubical epithelium, usually one cell thick, resting on a basement membrane and enclosing a lumen lined with a transparent membrane. Muscle fibres normally do not occur but are found in a few cases on the large arteries of Crustacea. The arteries generally have elastic properties: smaller arteries lack a visible lining membrane and may form networks anastomising on the surface or within the tissues of the organs they supply.

THE NEURAL ORGANS

The organs of the nervous system are the brain, the ganglia and the neuro-haemal organs. The brain and ganglia are special centres for the processing of incoming sensory information and correlating it with outgoing patterns of motor instruction, either already present or created in response to the incoming messages. The neuro-haemal organs are part of a special output mechanism of the nervous system.

In *Peripatus* (Fig. 4.25a) the brain consists of a nerve mass lying in the haemocoele dorsal and anterior to the fore-gut. The cell bodies form a layer over the surface of the brain, the axons and terminalizations a central core. Nerves connect the brain to the eyes, antennae, mouthparts, and anterior body wall. Posteriorly, a pair of large nerves arise, pass ventrally one on each side of the gut and turn posteriorly. Then, lying each in the haemocoele medial to the limb bases, they continue backwards and join dorsal to the hind-gut. In each segment behind the third, five transverse commisures unite the cords across the mid-line, and laterally, six to ten nerves spaced down the length of the cord supply the limbs, the body wall muscles and the sense organs. These additional neurons result in the cord being thicker in the segments than between the segments. These are the segmental ganglia and the intersegmental regions of the cord joining them are the connectives. The cell bodies lie in the dorsal, their axons in the ventral, part of the cord.

In other arthropods the following changes occur. The nerve cords and ganglia come to lie much closer together in the mid-line. The left and right ganglia of a pair fuse and the paired nature of the cord is only apparent from the paired connectives joining the ganglia. The cell bodies of the neurons become confined to the ganglia. The connectives contain only axons.

The detail of the processing of nerve information in its passage through a ganglion is primarily a matter of the kinds of neurons present and the patterns of contact each makes with the other. Within a ganglion many different types of neuron occur. No complete account of the structure and

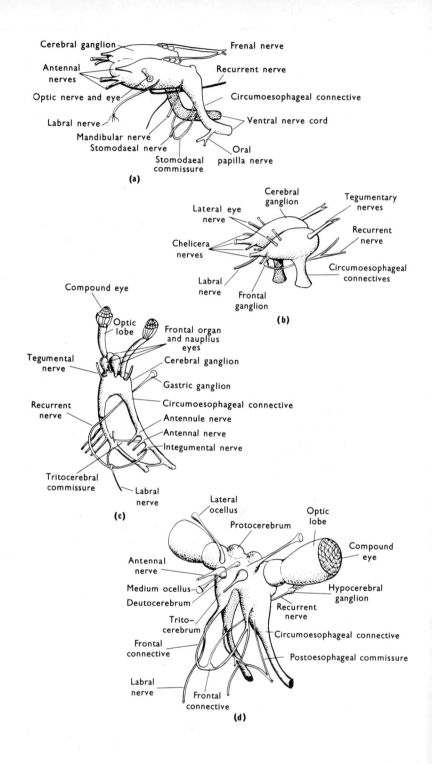

(a)

Cerebral ganglion
Antennal nerves
Optic nerve and eye
Labral nerve
Mandibular nerve
Stomodaeal nerve
Stomodaeal commissure
Frenal nerve
Recurrent nerve
Circumoesophageal connective
Ventral nerve cord
Oral papilla nerve

(b)

Lateral eye nerve
Chelicera nerves
Labral nerve
Cerebral ganglion
Frontal ganglion
Tegumentary nerves
Recurrent nerve
Circumoesophageal connectives

(c)

Compound eye
Optic lobe
Tegumental nerve
Recurrent nerve
Tritocerebral commissure
Labral nerve
Frontal organ and nauplius eyes
Cerebral ganglion
Gastric ganglion
Circumoesophageal connective
Antennule nerve
Antennal nerve
Integumental nerve

(d)

Lateral ocellus
Protocerebrum
Optic lobe
Antennal nerve
Medium ocellus
Deutocerebrum
Trito-cerebrum
Frontal connective
Labral nerve
Frontal connective
Compound eye
Hypocerebral ganglion
Recurrent nerve
Circumoesophageal connective
Postoesophageal commissure

function of a ganglion at this level exists and the knowledge available relates to a much coarser level. In any ganglion the axons of some of the neurons tend to lie parallel to one another for part of their length forming nerve tracts. The general function of some of these tracts, i.e. whether carrying sensory, integratory or motor information, is known. The cell bodies of the neurons tend to lie together in different parts of the ganglion; such a mass of cell bodies is often called a 'nucleus' and may be associated with an axon tract of known function. The terminal arborizations of the neurons associate together to form a welt of nerve tips and synaptic junctions where information is sorted and passed amongst the neurons: this is the neuropile.

The ventral nerve cord ganglia are dominated by the locomotory requirements of the animal. As the limbs evolve and the role of the body wall musculature in locomotion disappears, there is a need for more numerous instances of integration between muscles within the limbs, between the left and right limbs of a segment, and between pairs of limbs in successive segments. The ganglia reflect this need with the nerves arising closer together, increased organization within the ganglia and the occurrence of greater numbers of association neurons.

Within a ganglion, neurons are associated together to give the following general structure (Fig. 4.26): sensory nerves enter the ganglion ventro-laterally; the motor axons leave dorso-laterally. Fusing to give nerves of mixed function occurs just outside the ganglion. Fibres passing in connectives anteriorly and posteriorly lie on the dorsal and ventral regions of the ganglion on each side of the mid-line. Association fibres occupy the middle part and their cell bodies and those of the motor neurons lie in the lateral regions of the ganglion. In many cases the distribution of the axons of individual neurons is known. Some, perhaps the majority, terminate on the side of the ganglion at which they enter (ipsi-lateral side); others end on the opposite (contra-lateral side). Many do not end in the ganglion they enter but pass to more anterior or posterior ganglia, either ipsi- or contra-laterally.

The size of the ganglion is positively correlated with the complexity of the structure and function of the segment with which it is associated. For example, it is small in segments such as those of the insect abdomen which lack locomotory appendages, but it is large in the thoracic segments where locomotory activity is concentrated. In all groups of arthropods examples occur where ganglia of successive segments are fused together to form a large compact nerve mass. Usually the stable ganglion (that which keeps its segmental position), is the one associated with the most complex segment, the others moving forwards or backwards to fuse with it. Internally the

Fig. 4.25 Diagrams of the brains of some arthropods. (a) *Peripatus*. (b) Scorpion. (c) Primitive crustacean (*Triops*). (d) Insect (*Locusta*).

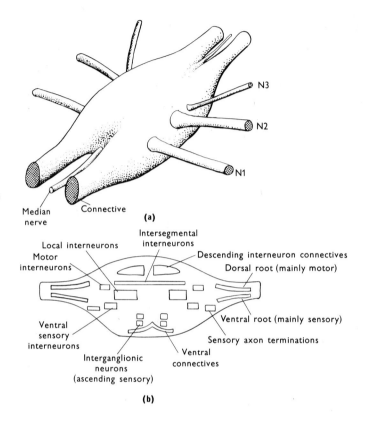

Fig. 4.26   Diagram of a ganglion from the ventral nerve cord. (a) External
appearance. (b) Section showing the distribution of the main tracts within the
ganglion. (Based on that given for a dragonfly larva by Zawarzin, 1924)

structure still shows the segmental arrangement of tracts and nuclei of the
individual ganglia which compose it.

The external form of the brain (Fig. 4.25) differs greatly from one group
to another, its shape being controlled partly by the surrounding anatomy
of the head and partly by the dominance of different sensory inflows in the
different species. The head exoskeleton does not fit closely round the brain
as does the skull in vertebrates and brain shape is only slightly correlated
with the head form.

The dominance of a sensory inflow on brain form is expressed through
the increased numbers of neurons associated with its greater capacity. In
forms where the optic inflow is small, i.e. those possessing ocelli, only the

ocellar nerve where it enters the brain has at its root an associated mass of neuropile of quite modest dimensions, producing no more than a slight swelling on the brain surface at this place. Most arachnid eyes fall into this class as do the ocelli of insects. Where compound eyes are present, well-developed large optic lobes occur in the brain. These have the external form of cones of brain tissue arising from the protocerebrum and ending at the base of the compound eye: the narrow part of the cone is adjacent to the protocerebrum, the broad part to the eye. Internally the optic lobe (Fig. 4.27) has a complex structure. The axons from the retinula cells where they enter the brain lie in bundles and tracts for a very short distance; then occurs a region of neuropile called the lamina where many fibres end in complex synaptic patterns with other neurons; another region of axon tracts follows and again a complex pattern of interchange gives rise to a region of distinctive neuropile, the medulla. This is repeated again, the inner pattern being the lobula. The fibres leaving the lobula pass to other parts of the brain and ventral nervous system. This description applies particularly to the insects. A similar pattern occurs in Crustacea but the medulla is not a single region spanning the width of the optic lobe, but two regions of lesser diameter, one slightly behind the other. Two regions are present in Lower Crustacea and Myriapoda.

In arthropods with antennae a distinctive middle region of the brain, the deutocerebrum, is present. It contains the motor centre of antennal movement and performs the analysis of antennal sensory information. In Crustacea chemoreception by antennal organs is important and large swellings, the 'olfactory' lobes, are associated with the root of the antennal nerve.

Throughout the different groups of arthropods the internal structure of the brain (Fig. 4.27) shows certain tracts and association centres which, while not necessarily homologous, are sufficiently similar in position and appearance to be judged analogous and to bear a common name. The most important of these are:

1. The central body, a medium mass of neuropile lying in the proto-cerebrum.

2. The protocerebral bridge, a tract of axons connecting the lateral parts of the brain across the mid-line in front of the central body.

3. The corpora pedunculata (mushroom bodies), a pair of bodies situated one on each side of the mid-line in the protocerebrum. They consist of a region of unipolar neurons with large nuclei in densely stained cell bodies arranged in the form of a circular cap indented on the posterior side from which a stalk of dense neuropile arises from several roots and passes back into the brain, turning towards its fellow in the mid-line distally. This description is founded upon insect material. In Crustacea similar structures, the hemi-ellipsoid bodies occur.

Evidence suggests that these bodies and tracts are concerned with integration of nerve information which relates sensory input to motor output

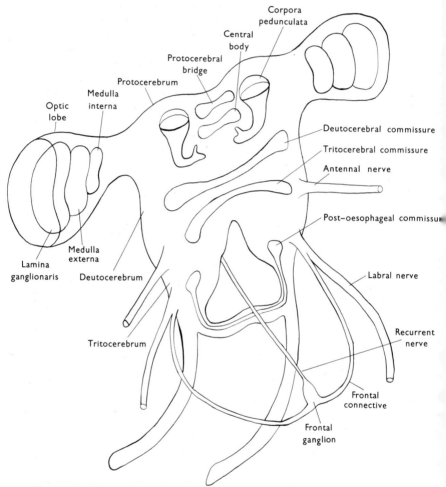

Fig. 4.27   Diagram, based on the brain of an insect, showing the position of the major association and optic ganglia within the brain of an arthropod.

and controls patterns of nerve activity. The size of the central body varies greatly in different arthropods, being for example, small in phalangids and myriapods, of moderate size in many spiders, crustaceans and insects, large in some insects and enormous in web-building spiders. Only in spiders is large size positively correlated with complexity of behaviour. In social insects the central body is small. The protocerebral bridge has not been assigned a definite function. The corpora pedunculata also vary greatly in size and details of form in different arthropod groups. They are enormous in *Limulus* and in *Periplaneta*. At our present level of knowledge its functions are various. In bees and termites it is of large size and is positively correlated with complex behaviour, but this is not so with the ants, and it is large in the wingless blind insect *Japyx* whose behaviour patterns are simple.

In the Crustacea it appears to be concerned with relationships between information from the olfactory lobes and some of the reflex activities of the animal. In spiders it is concerned with integration of optic input. The region is well developed in jumping and wandering spiders with well-developed eyes, less so in sedentary ones. It is negatively correlated with the size of the central body in web-building spiders and completely absent from the brains of some of them (*Tegenaria*).

Neurosecretory cells are to be found in the brain, ganglia of the ventral nerve cord, and the stomatogastric ganglia of all Arthropoda. The secretion is conveyed from the perikaryon where it is synthesized, through the axon, to the terminal arborizations where it may be stored. Release of the product from the cell occurs at the terminal dendrites. The axons may conduct the secretion to its target organ as occurs in the neurosecretory cells supplying the corpora allata. In many instances however, the secretion is released into the haemocoele and, at the site of release, special relationships occur between the terminal dendrites of the neurosecretory cells and the haemocoele, forming the neuro-haemal organs.

In its simplest form a neuro-haemal organ is a thickened disc of neurolemma with terminal dendrites containing neurosecretion against its inner surface, and a blood sinus against its outer one. This structure may become more elaborate, the neurolemma becoming cup-shaped forming channels and making a more complex and greater surface area of contact between the haemolymph and the neurosecretory terminal dendrites. The axons filled with neurosecretory material may become orientated perpendicularly to the free surface of the organ, interspersed with non-glandular cells, giving the tissue a pallisade appearance. Material may be stored in the terminal arborizations which then form an extensive pad, or it may be released and no appreciable store can be detected. Organs at this level of structure are: possibly the infracerebral organs of *Peripatus*; the sinus gland, post-commisural organs, and pericardial organs of Crustacea; the cerebral gland of Chilopoda.

In addition to cells supporting the terminal arborizations of the neurosecretory cells and facilitating their contact with the haemolymph, some neuro-haemal organs contain cells which produce either an intrinsic secretion of their own, or modify the product of the neurosecretory cell. These organs are the corpora cardiaca of Insecta and the neuro-haemal organs of the Arachnida.

5

The Anatomical Systems of Arthropods

INTRODUCTION

The organs are parts of systems, a system being an anatomical entity whose structure and disposition within the body can be correlated with the performance of a set of functions, one of which usually predominates. A system is named after its dominant function. In arthropods the following systems can be recognized:

1. The muscular system.
2. The skeletal system formed from the integument.
3. The nervous system.
4. The digestive system.
5. The respiratory system.
6. The blood-vascular system.
7. The coelomic organ system.
8. The reproductive system.

This account omits an excretory system since excretion of nitrogenous waste is carried out by tissues which are parts of organs related to other systems: e.g. in Crustacea ammonia is excreted through the gills; in Insecta uric acid, or in Arachnida, allantoin and guanine, from the Malpighian tubules which are derivatives of the gut.

Not every tissue or organ can be assigned to a clearly defined anatomical system. The integumental glands with their many diverse functions do not act as an integrated whole. Similarly, the fat body of insects, important though it is, cannot be associated with other organs to form an anatomically definable system.

THE MUSCULAR SYSTEM

The muscles can be classified into those associated with the body wall and its derivatives, and those with the visceral organs. Both are formed from smooth muscles in the Onychophora and from striated muscles in all other arthropods. The muscles associated with the body wall can be grouped into two further sub-categories: those moving the segments relative to each other, and those moving the limbs. The latter consists of the extrinsic muscles which move the limbs relative to the body, and the former of the intrinsic muscles which move the podomeres relative to each other.

The segmental muscles (Fig. 5.1)

These arise from the circular and longitudinal muscle coats of the body wall of the primitive non-segmented animal. With the development of segmentation the muscles of these coats were formed from paired serially arranged mesodermal tissue blocks. The muscle fibres of the complete coat did not have the segmental limits of their initial formation, but extended through several segments. A muscle coat of this type is found in Onychophora where outer circular and inner diagonal muscles form a continuous sheath beneath the basement membrane. Such coats are associated with the lack of a firm skeleton and considerable ability to change the body shape.

The modification of this primitive condition to that found in present-day arthropods is the result of the interaction of two forces: the increased 'segmentalization' of the muscle coats and the development of an exoskeleton. Increased segmentalization organizes the muscle coat into a number of circular and longitudinal muscle blocks separated from each other by intersegmental connective tissue in sequence down the length of the body. Movement of one body segment relative to the next is brought about by the longitudinal muscles which arise from one intersegmental disc and are inserted upon the next. It is doubtful if this 'ideal' circumstance ever actually occurred in arthropods, but it forms a useful framework against which to visualize the effects of the development of an exoskeleton.

With the development of an exoskeleton the circular muscle layer (Fig. 5.1) is greatly reduced. The longitudinal layer is less reduced and consists of a thin band of evenly spaced muscle fibres forming a series of segmentally placed cylinders lying beneath the epidermis. The dorsal and ventral parts of such cylinders may be thicker than the lateral parts. In the wing-bearing segments of insects or the abdominal segments of decapod Crustacea, the dorsal longitudinal bands increase in size to form powerful muscles. Where the exoskeleton is greatly reduced the muscle bands increase in thickness to form again longitudinal, circular and often diagonal muscle coats lying beneath the epidermis. The final pattern assumed by

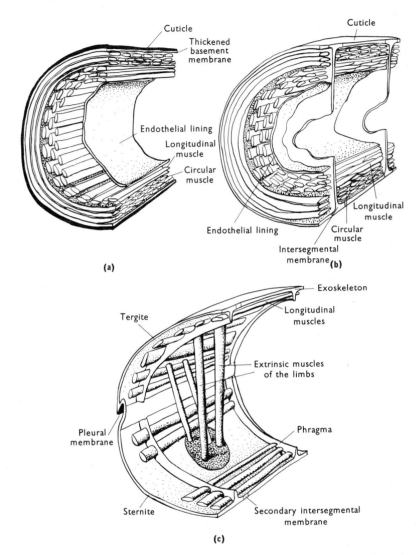

Fig. 5.1 Diagram showing the development of the segmental muscles of arthropods. (**a**) Muscular wall of a primitive non-segmented, limbless animal showing outer circular and inner longitudinal muscle coats. (**b**) Muscular wall in a segmented limbless animal, showing the muscles broken into segmental groups. (**c**) Muscular system in arthropods possessing an exoskeleton and limbs. The circular muscles have practically vanished, the longitudinal coat is greatly reduced, and extrinsic muscles have developed for moving the limbs.

the muscles in any arthropod is dictated by the requirements of movement of the animal, and these requirements can override the segmental organization of the animal. Muscles can cross the body vertically, diagonally or horizontally to meet the needs of the animal.

Few generalizations can be made about the muscles (Fig. 5.1) which operate the appendages of arthropods. The simplest basic picture is that of muscles originating in one podomere being inserted in the next, moving the distal one relative to the proximal one. This pattern is repeated throughout the succession of podomeres which make up the limb. In the absence of special articulations between the podomeres a large number of fibres form a cylinder capable of moving the distal podite in any direction and holding it in any position relative to the proximal one. With the development of articulations and the limiting of the field of movement of the distal podomere, the muscle ring can also be reduced. The reduction may simply leave two muscles, an extensor muscle which straightens the limb and a depressor muscle which bends it. In some cases, the legs of spiders, only the depressor muscle is present, extension being carried out by haemolymph pressure.

The musculature of the basal podites must be powerful enough to support and move the weight of the distal ones. The origins of the muscles move from the tip of the podite towards its base and are inserted on to the distal podite by thin tendons of much lighter mass. In general, all that can be done to get the muscle mass as close to the base of the limb as possible while maintaining the proper functions and movements of the limb is to be found in the limbs of the more advanced arthropods.

The power of a muscle varies in proportion with its cross-sectional area, so that short wide muscles occur where powerful relatively slow movements are required, and long thin ones where quick relatively weak movements are sufficient, the weakness being compensated by an increase in the number of muscles present. For example, in slow-moving but powerful millipedes two muscles may be sufficient to move the limb relative to the body wall, but 4–5 muscles occur where fast movements are required.

THE SKELETAL SYSTEM (Fig. 5.2)

In the Onychophora the skeletal system is a thick tough basement membrane lying beneath the epidermal cells. The cuticle forms a thin, inelastic, but very flexible outer covering of the epidermis and has no skeletal function. In all other arthropods the cuticle is thickened and strengthened so that it is capable of acting as an exoskeleton. The exoskeleton consists of a number of hard rigid plates, sclerites, between which are softer, more flexible regions which permit the movement of one sclerite relative to the next.

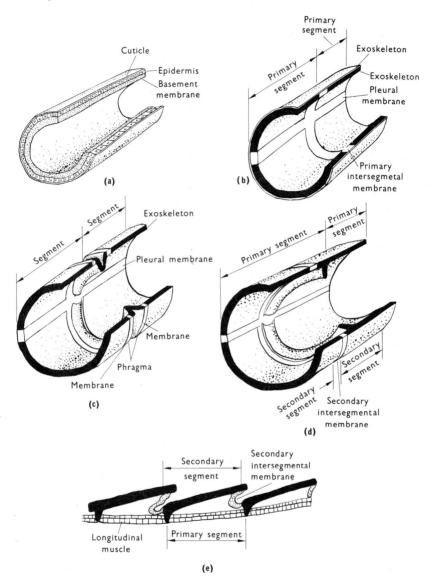

Fig. 5.2 The skeletal system. (**a**) Skeletal system of a simple arthropod. (**b**) Diagram of simple arrangement of skeletal plates and soft membranes. (**c**) Development of 'intersegmental' skeletal plates phragmata for insertions of the segmental muscle blocks. (**d**) Fusion of phragma with either preceding or succeeding segmental plate. (**e**) Longitudinal section of tergites of segments showing development of secondary segmentation.

The dominant forces acting at the system level which dictate the pattern of sclerites and membranes are those which bring about the correct mechanical relationships between the primitive muscular body wall and the sclerites to ensure efficient movement of parts of the skeleton; and the changes brought about by the locomotory adaptations of the animal.

The simplest skeleton (Fig. 5.2b) consists of segmentally arranged dorsal plates, the tergites, and ventral plates, the sternites, having laterally a flexible pleural membrane. Occupying an intersegmental position between one set of plates and the next is a flexible annulus of cuticle, the intersegmental membrane. The basal podite of the limb, the precoxa, arises from the flexible pleural membrane between the tergite and the sternite. At an early stage this podite formed the pleural sclerites lying anteriorly and posteriorly to the limb base and providing articulations and a base for the muscles which operate the basal limb podite (coxa) in all modern arthropods.

Within the segmental membrane dorsal and ventral sclerotic plates developed. These form by a deep infolding of the epidermis and the secretion of a sclerotic plate (phragma) between the opposing surfaces of the fold. Only a narrow sclerite is visible externally and is separated from the tergites or sternites of the preceding and succeeding segments by flexible membranes. The longitudinal muscles become inserted on to the surfaces of the phragmata, spanning the distance between and moving the plates relative to each other. Transmission of this movement to the tergite or sternite of a segment is effected by a fusion between either the preceding or the succeeding segmental plate and the intersegmental sclerite, the remaining membrane enlarging and forming the flexible annulus between one set of sclerotic plates and the next. The boundaries of these plates do not now correspond to the original intersegmental boundaries but lie in either the posterior region or the anterior region of the segment. A new secondary segmentation is formed; each secondary segment consisting of the intersegmental region and a portion of a segment of the original segmental organization of the animal. Secondary segmentation is a basic feature of arthropod skeletons and results from the relationships between the primitively segmented longitudinal and circular muscular body wall and the surface exoskeleton of the animal.

Secondary segmentation may be more complicated. Tergites may fuse with the anterior intersegmental plates and sternites remain unfused or fuse with the posterior intersegmental plate. The flexible annulus marking the visible segmental boundaries will then follow a complex transsegmental path. Finally, fusion between succeeding segmental and intersegmental plates may produce rigid sclerotic structures extending over several segments, in which all external trace of segmentation may be obliterated, as in the head capsule.

The changes that lead to locomotor efficiency

The limbs of arthropods both support the body above the substratum and effect locomotion. In *Peripatus* the limbs arise from the ventro-lateral aspect of the body. Lengthening of these limbs would raise the animal further above the substratum and eventually reach a point where the animal would be so elevated on its limbs that it would be unstable. In arthropods lengthening of the limb is always associated with its bending to approximately an S-shape. The limb originates ventro-laterally, bends horizontally, then vertically and finally downwards to the ground. The body of the arthropod is 'hung' from its limbs and the increased length of the limb does not cause instability of posture (Fig. 5.3a). The only arthropods which stand up on straight limbs are inhabitants of very sheltered places, or, since an increase in body weight will also increase stability, are those of relatively

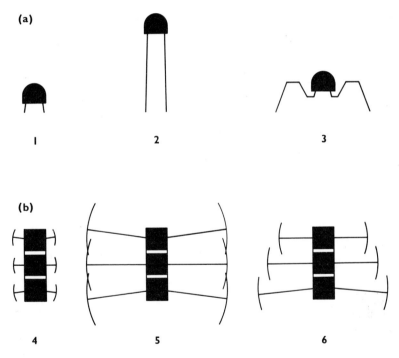

Fig. 5.3 Problems of increased limb length. (a) *Instability.* **1** Primitive stable configuration, **2** unstable configuration with long limbs, **3** stable configuration with long limbs. (b) *Overstepping.* **4** Short limbs, no overstepping; **5** long limbs, marked overstepping when the full swing of the limb is used; **6** long limbs, no overstepping.

large body size, like the racing crabs *Ocypoda* which run with their nearly straight legs raising the body well clear of the ground.

Another problem associated with increased limb length is the possibility that the field of movement of one limb may overlap that of its predecessor or be overlapped by its successor in such a manner that the efficient movement of one is hindered by the movement of the others. This is avoided in two ways:

1. In centipedes, and perhaps primitively in other arthropods, the limb ends in an acute tip which contrasts with the blunt pads of *Peripatus*; the tip can be placed down alongside the other limbs, thus getting the maximum stride without overlap.

2. The successive limbs may be of different length so that their fields of movement lie in different planes and even with overlap one limb does not interfere with the other (Fig. 5.3b). This can be seen in *Scutigera*, where limbs of each posterior segment are longer than the one succeeding it, and in insects where each limb pair has a different length. Another solution is seen in crabs where their sideways movement caused by limbs moving towards and away from the mid-line prevents overlap. This elimination of overlap enables the animal to walk in a greater number of ways than would otherwise be possible.

Peripatus can change the manner of its walking according to the speed it needs to go. This change is effected by (1) quickening the pace, (2) employing a gait with a faster pattern (shorter back-stroke, long fore-stroke), (3) using a larger angle of swing of the leg.

The duration of the pace is of little importance in *Peripatus* but becomes important in other arthropods where it contributes about 38–85% of the speed increase in *Polydesmus*, *Lithobius*, *Forficula*, and a lycosid spider.

Change in the gait contributes something like 47–75% to the increase of speed in *Peripatus* but only 8–25% in arthropod verra. It contributes most in those animals where the number of legs is reduced and fields of movement are much greater, i.e. some 5% in *Polyxenus*, 25% in the beetle *Tenebrio* and 30% in Hemiptera.

The change in the angle of swing contributes from 31–57% to the change of speed. In a spider and *Scolopendromorpha* the highest speeds are attained predominantly by a decrease in pace duration, in *Lithobius* and *Forficula* by an increase in the angle of swing.

The legs of a segment may be in phase, i.e. left and right legs in the same stage of the movement cycle, or they may be in different stages, usually opposite. The former situation is associated with powerful pulling or pushing, the latter with faster movement. In *Peripatus* both cases occur in the same animal in different types of locomotion but in arthropoda verra they are usually fixed for a given type of animal.

With an increased length of leg and with legs of a segment out of phase there is a tendency for the body to undulate when it is moving forward. The usual movements of an animal are neither very fast nor very slow and are generally performed with great efficiency. When, however, the animal has to go fast, the faults of the system occur and it is under these conditions that the different limitations of body form appear. *Lithobius* in normal movement has a straight body, but when startled into moving fast by being exposed when a stone is overturned, it tries to run quickly and characteristic undulations appear. These undulations place a severe limitation on the efficient use of limbs in locomotion and their removal has dictated to a large extent the body form of the different groups of arthropods.

In short-legged forms these may be countered by intersegmental musculature holding the body in a straight line, thus countering the lateral pressures produced. This will suffice even a long-legged animal if walking slowly. In many cases the muscles may extend over several segments. The intersegmental membranes may be staggered by alternately long and short terga. In *Ligia* short wide segments tend to reduce lateral deviations.

A reduction in the number of legs to a few pairs borne on adjacent segments, together with a fusion of the body segments concerned, effectively prevents any undulations arising from this cause. In arachnids four pairs of walking limbs occur on the prosoma, and *Galeodes*, one of the fastest of land arthropods, habitually moves only on three pairs. The three pairs of legs found in insects, larval mites and millipedes offer a convenient small number. Less than two pairs of legs, except for climbing, are not used in arthropods.

When limbs become reduced to a few pairs these are always at the anterior end associated with the need to give support to the head during feeding. Those segments often atrophy, as they are not now contributing to locomotion.

THE NERVOUS SYSTEM (Fig. 5.4)

The basic pattern of the nervous system is an anterior dorsal brain from which a pair of nerve cords pass ventrally one on each side of the gut and then, lying in the haemocoele close to the ventral surface, continue posteriorly to the extremity of the body. Commisures (a number per segment) join the cords across the mid-ventral line giving a characteristic ladder-like appearance to the system. Nerves arising laterally from the cords pass to and from the organs in each segment. The stomatogastric nervous system which supplies the gut has its roots in nerves arising from the brain and anterior cord. The major anatomical modifications of the nervous system result from a concentration of neural tissues to give a

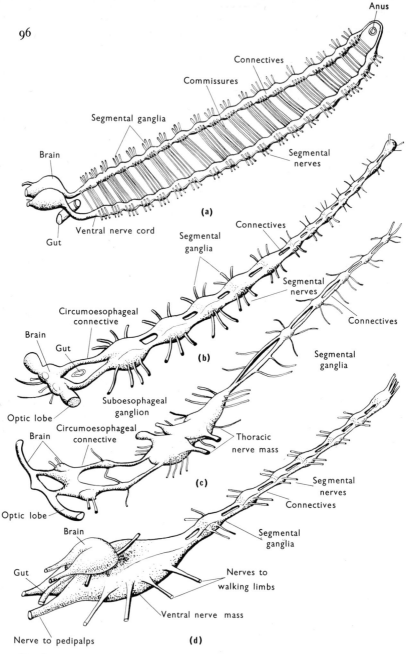

Fig. 5.4 Diagram of two main types of nervous system. (**a**) Type I, *Peripatus*. (**b**) Type III, *Locusta*. (**c**) Type III, *Triops*. (**d**) Type III, Scorpion.

single nerve mass in which little or no trace of the early ladder-like pattern can be seen. Four steps in this direction can be distinguished:

1. The primitive open ladder-like structure found in *Peripatus* (Fig. 5.4a) and taken to represent the arthropod nervous system at its most elementary stage of organization.
2. The development of a pair of ganglia in each segment and their fusion in the mid-line (Fig. 5.4b). The ventral nerve cords now lie close to each other in the mid-ventral line, communication between them being restricted to the ganglia to which the nerves to and from the segment organs are now concentrated. This stage is widespread throughout the primitive members of all groups.
3. The paired ganglia of successive segments fuse together to produce larger and more complex nerve masses (Fig. 5.4c, d). The patterns of fusion that can occur in a chain of ganglia are numerous and related to the external form of the animal. As an end product of many evolutionary lines all the ventral ganglia fuse to form a common mass, the nervous system then consisting of a dorsal brain, circumoesophageal nerve cords and a ventral ganglion from which nerves pass to all the post-oral segments, e.g. crabs and spiders. In insects, a brain, circumoesophageal connectives, suboesophageal ganglion and ventral thoracic ganglia form the most concentrated level normally found.
4. In many ticks, the brain and ventral ganglia have fused into a single mass, all external trace of segmentation having been lost to the nervous system.

The functions of the nervous system depend almost entirely upon its organization at the cellular, tissue and organ grades of structure and function. The pattern of nerves and ganglia which constitute the system is a result of the growth of axons out from the nerve cords and from the ganglia to the organs they innervate and of axons from sensory organs in towards them. Little is known of the mechanisms which dictate the paths the axons follow and while within a species a general pattern can be determined, there are considerable differences between individuals. Major differences can often be correlated with sex and with age but most of the variation is simply an expression of the uniqueness of the individual animal.

The welding together of the nervous system depends not only upon the chain-like nature of neurons but also on the presence of neurons which extend through all or large parts of the system. In *Limulus* neurons occur in which the cell body is situated in the protocerebrum and the axon has arborizations in each succeeding segment. In *Locusta* axons from the retinula cells of the eye pass to the thoracic cord without synapse and, in the scorpion, giant fibres control the sting and connect its reactions through other giant fibres to the pedipalps and walking legs. These fibres, which co-ordinate the leap, thrust and attack of the animal, transcend segmental boundaries and integrate activity in remote parts of the nervous system.

The peripheral nerves may also cross segmental boundaries. Many muscles in myriapods are supplied by nerves originating outside the muscle's segment and, in Insecta, the spiracles are innervated from the ganglion immediately anterior to them.

The anatomical pattern of the system, whilst basically imprinted with the segmental organization of the animal, carries important elements which override this and indeed must do so in order to perform its functions of integrating body activities.

THE DIGESTIVE SYSTEM (Fig. 5.5)

The details of this system have already been sufficiently discussed in Chapter IV.

THE RESPIRATORY SYSTEM (Fig. 5.6)

All respiratory systems are concerned with the entry of oxygen into the body and the exit of carbon dioxide from it. The degree to which the respiratory system actually conveys the gases to and from the tissues where they are used gives the basis for distinguishing between two main types of system in arthropods.

Where gills and lung books occur, the function of the respiratory system is to convey oxygen from the external medium to the pigment in the blood and carbon dioxide from the blood to the outside. In addition to the gills, other structures designed to facilitate the flow of water over them and to keep them clean make up the respiratory system. The gills themselves are capable of some movement but the area of water they sweep is small, and unless it is renewed by movement either of the animal or of the water, enough oxygen is not available. The gills are delicate structures and are often enclosed in a cavity formed by a fold in the body wall or are placed at the base of an appendage whose shape can protect them. Water flow over the gills is produced by movement of appendages, either by rhythmical beat of the appendage bearing the gills or by a rapid movement of a single appendage placed at the exit of the chamber and which bales the water out of the cavity. Water entry may be through one or several openings; within the cavity water flow is directed over the gills by contours of the cavity walls.

Particulate matter in the water is filtered out from the respiratory current by pads of hair and bristles across the inhalent openings. Even so, minute matter may still enter and settle out on the gill surface; this is removed by the action of a number of gill rakers, these being long processes (exopodites)

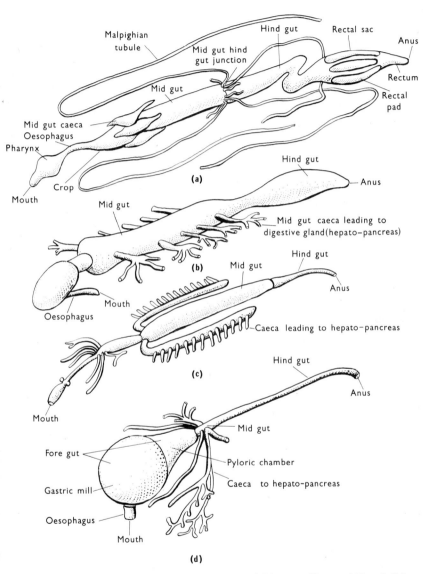

Fig. 5.5 Diagram of the gut type of arthropods, (**a**) Insecta (*Locusta*) Type I. (**b**) Merostomata (*Limulus*) Type II. (**c**) Arachnida (*Galeodes*) Type II. (**d**) Crustacea (*Cancer*) Type II.

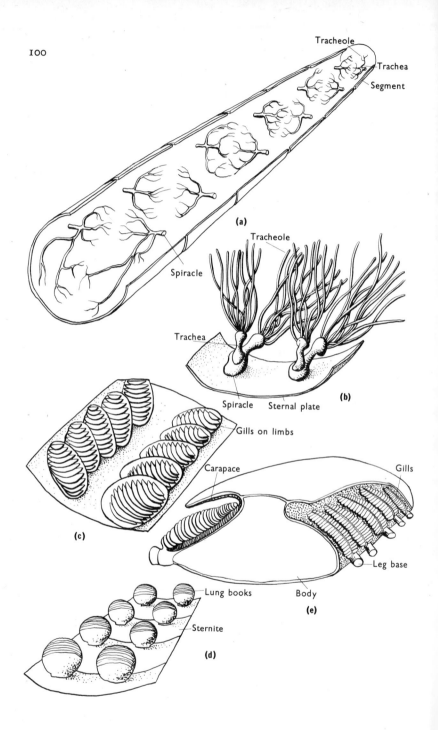

(a)

Tracheole

Trachea

Segment

Spiracle

(b)

Tracheole

Trachea

Spiracle   Sternal plate

(c)

Gills on limbs

Carapace

Gills

(e)

Leg base

Body

Lung books

Sternite

(d)

which sweep backwards and forwards over the gill surface, combing and cleaning them.

In lung books air is pumped in and out of the lung book cavity by movement of the wall. The aperture to the lung books is narrow to prevent entry of dust and to restrict water loss: it may be completely closed by valves. Lung books are always paired and one or more pairs may be present, spiders having four pairs and some scorpions six pairs.

The second main type of respiratory system is composed of spiracles, tracheae, air sacs and tracheoles, air being conveyed from the environment directly to the tissues where it is used and carbon dioxide directly away from them.

The most primitive system occurs in *Peripatus* where something like 1500 spiracles spread over the body surface constitute the respiratory system. Each appears to respond independently of the others to the condition of the tissues which it supplies.

In all other tracheal systems there is a great reduction in the number of spiracles and a great increase in the amount of tissue each supplies. Whilst typically there is a pair of spiracles leading to a branched trachea ending in tracheoles to each trunk segment, single dorsal openings occur in the centipede *Scutigera*, a single pair of spiracles on the membrane between the head and prothorax in Collembola (*Sminthurus*). Functional spiracular openings are often restricted to a few trunk segments or to only one.

The presence of air sacs, which are often thin-walled and inflatable, permits changes in body volume by allowing the increased or decreased size to be taken up by corresponding changes in the air content of the sac. Regular movements produced by movement of the tergites and sternites give regular ventilation to the tracheal system and shorten the diffusion path between the outside air and the tracheolar openings.

Air circulation through the tracheal system results from fusion between tracheae of succeeding segments and the left and right sides of the same segment, together with the development and integration of the movements of the spiracular valves guarding the tracheal openings. The expansion of abdominal volume is correlated with the opening of two or three pairs of anterior spiracular valves, air being sucked into the expanding air sacs: the anterior valves close, contraction begins, air pressure builds up in the

Fig. 5.6 Diagram of the types of respiratory systems. (a) Primitive tracheal system, one pair of tracheal trees per segment (Insecta, Myriapoda). (b) Tracheal system of Arachnida (*Ricinoides*), one pair of tracheal trees from one segment only. (c) Gills from *Limulus*, five pairs of gills on the posterior surface of five pairs of appendages. (d) Lung book system of Arachnida (Scorpion), four pairs of lung books, two per segment on the sternites of the opisthosoma segments. (e) Gill system of Crustacea (Decapoda), series of gills projecting from the thoracic walls and base of thoracic limbs, often lying in a chamber formed by the carapace.

tracheal system, the posterior valves open and air is expelled through the posterior spiracles. This regular breathing occurs in nearly all large active insects.

The distribution of air to those tissues that require it still depends upon the osmotic changes occurring in the tissues themselves, just as it did in *Peripatus.*

In small arthropods the diffusion of gases through the integument is adequate to meet oxygen requirements and no special respiratory organs or system are present.

THE BLOOD VASCULAR SYSTEM (Fig. 5.7)

The main function of the blood vascular system is to produce a circulation of blood around the body and amongst the tissues and organs. The blood can then act as a transport medium to convey materials from one part of the body to another.

The anatomy of the system within the arthropods shows a range of variation between and including two major Types. In the first Type (arterial), the heart discharges blood into the arteries which convey it to organs where it comes into contact with the tissues from which blood returns to the pericardial sinus and thence to the heart through the haemocoele. In the second Type (aortic), the arterial system is greatly reduced, often forming only an aorta directing blood to the head; from here blood flows through the haemocoelic spaces to the organs and thence back through other parts of the haemocoele to the pericardial cavity and the heart. In both Types the heart may be either tubular and peristaltic or compact and beating.

In the first Type the direction of blood flow is definite and the total volume of blood can be directed to or passed through an organ. In the second Type the blood flow is less constrained; it is effected by local movement of organs in the haemocoele and the entire blood volume cannot be directed to flow past an organ, although in time it probably does so.

Between these two extremes are many intermediate stages, but in general, Type I is found in the larger Arachnida, Crustacea and Myriapoda, Type II in Onychophora and Insecta. Smaller arthropods tend towards Type II circulation but their small blood volume and large heart capacity make the total circulation time quite short.

With regard to the transport of materials by the haemolymph, the nature of these materials bears some relationship to the type of circulation present. Where respiratory pigment is present in the haemolymph, swift circulation must occur since oxygen is required in continuous supply and the storage capacity of the tissues and the carrying capacity in simple solution by the haemolymph is not adequate to allow slow transport. In the absence of a

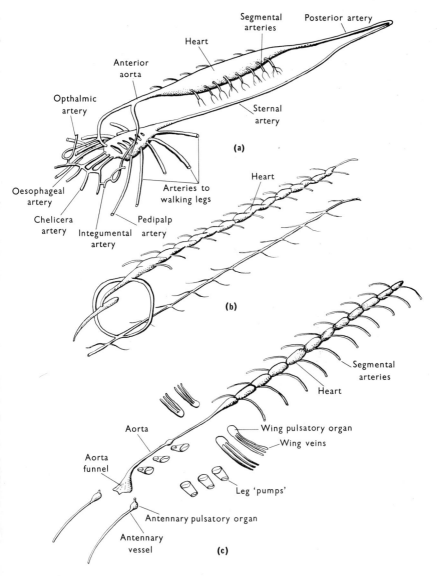

Fig. 5.7 Types of blood vascular system within the arthropods, (a) Type I Arachnida (Scorpion) very complete arterial system. (b) Type I Crustacea (*Squilla*) arterial system much reduced. (c) Type II Insecta (*Periplaneta*) arterial system greatly reduced, subsidiary pumping organs developed to circulate blood through antennae, limbs and wings.

carrier function by the haemolymph, slow turnover rates of other materials are met with less fast circulation.

In Type II systems local subsidiary circulations can occur, as in the wings, antennae or legs: otherwise the ventral-dorsal movement of blood might override the anterior-posterior one. Thus local transport can occur without overall circulation of the blood and turnover times in these local systems are less than they would be if total circulation were necessary.

These local circulations may well be facilitated where the distance between organs is reduced; one often finds that in Type II circulation the organs of the body are branched and ramify throughout the haemocoele so that linear distance between them is short and a slow seepage of blood is adequate for transport.

THE COELOMIC SYSTEM

This system, concerned mainly with water and ion regulation in the animal by controlling the output of these substances from the body, is most fully represented in *Peripatus*, where a pair of coelomic organs occur in each body segment behind the slime papillae, excepting that bearing the reproductive apertures. Some 18 to 30 pairs of such organs may be present.

In other arthropods this number is greatly reduced and in some groups they are absent. Two pairs of such organs may survive on adjacent segments in some Arachnida; their ducts join between segments and have but a single pair of openings. In Crustacea one pair, the antennal glands, serve their function in the adult and another pair, the maxillary glands, in the pre-adult stages. In Insecta a single maxillary pair serve as salivary glands in some Apterygota, pumping fluid down on to the surface that is being rasped away for food, the fluid then being swallowed to convey the food to the gut. In pterygote insects coelomic organs are absent.

The reproductive system (Figs. 5.8, 5.9)

While embryological studies show that the tissues of both the ovaries and testes receive contributions from the mesoderm of a number of segments, only a single pair of these organs is present in the adult. In a few groups, notably the barnacles, the animals are hermaphrodite but usually either testes or ovaries are present. Primitively each gonad of a pair opens separately to the exterior through a pore on the ventral side just medial to the appendage base. In males and females of the same species the gonopores may open on different segments. The segments bearing the gonopores are different in different groups. In Diplopoda they are the anterior trunk segments; in Arachnida and Crustacea, the middle trunk segments; and in Chilopoda and Insecta, the posterior trunk segments.

In the male the sperm matures in the testis and is passed to the gonoduct where it may be stored in its swollen lumen, the seminal vesicle, before passing to the gonopore. The sperm appear hardly ever to be broadcast at random into the environment but to be enclosed in a spermatophore produced either by the glandular lining of the gonoduct or by special ectodermal glands, the accessory glands, developed from the immediate posterior segments. The spermatophore containing the sperm is transferred to the female in a number of different ways. Rarely the spermatophore

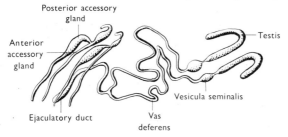

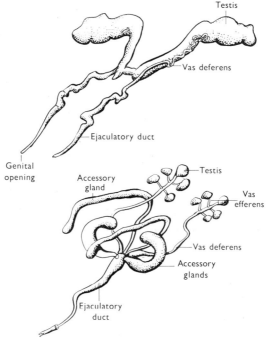

Fig. 5.8   The reproductive system of some male arthropods. (a) Onychophora, *Peripatus*. (b) Crustacea, *Cancer*. (c) Insecta, *Tenebrio*.

may be plastered on to the integument of the animal. Then the area of integument underlying the spermatophore and the spermatophore wall at this point is eroded away and sperm enters the haemocoele, eventually reaching the female reproductive system. More often the spermatophore may be transferred to a specially modified appendage, as in spiders, and the appendage inserted into the female gonopore, carrying the sperm with it. The appendages may be those of the gonopore segment, and are then

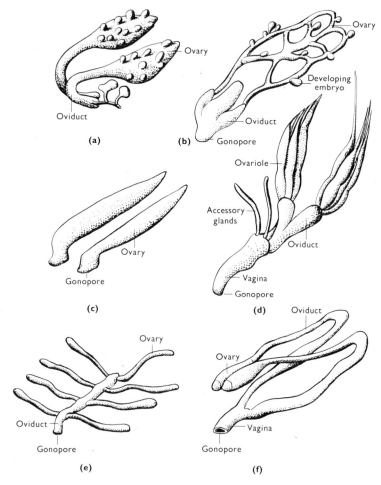

Fig. 5.9   Diagrams of the reproductive system, female. (a) Spider. (b) Scorpion. (c) Crustacea. (d) Insect. (e) Pycnogonida. (f) *Peripatus.*

modified to form tubes which carry the sperm directly into the female system; under these conditions the spermatophore may be absent.

In the female, the eggs produced in the ovary pass to the oviduct where they come into contact with sperm ejected from its store in the spermatheca. The eggs then pass down the oviduct and out of the animal. Appendages may be modified to hold the eggs until they hatch.

In many cases the eggs are enclosed in a shell or chorion produced by the follicular cells within the ovary. The chorion is an elaborate structure designed to protect the eggs against environmental hazards. Special pores, the micropyles, occur in the chorion through which sperm can penetrate when the egg passes the spermatheca.

The gonoducts in many arthropods do not retain their separate openings but share a common opening in the mid-ventral region of their segment. In the female the common oviduct so produced can be extended posteriorly over one or two segments by grooves appearing in their mid-line, eventually becoming tubes carrying the gonopore more posteriorly.

Although the prime birth product of the majority of arthropods is an egg, some members of all groups retain the egg within the oviduct until it is developed and thus they give birth to live young. Where this happens modifications occur which permit passage of food from the mother to the developing young.

6

The Organization of the Arthropod Body

INTRODUCTION

The last stage in describing the body design of arthropods is accomplished in three sections: the relationships between systems and a method of expressing these in diagrammatic form; the segmental structure and general body form of the major groups of arthropods; a synthesis to embrace all the information so far presented. In the matrix on p. 234–245 is set out the essential structural and functional features which must be present in any arthropod. In each set is listed the main alternative conditions of each feature that are known to occur in the group. By selecting one of these alternatives from each set a description of an arthropod can be assembled which gives the major features of its design. Once this major pattern has been selected, it can be further elaborated by reference to the detailed descriptions of systems, organs, tissues and cells already given. The principles already discussed at each of these levels will guide selection where alternatives are possible or help in the assembling of systems, organs and tissues where the actual examples given do not express the structure and function required. It is thus possible to carry the description of an arthropod to different levels, and in greater and lesser detail, depending upon requirements.

The arthropod design so described should be structurally and functionally sound, but incomplete; since data on life history, control mechanisms, adaptations and evolution to be given in later chapters still has to be taken into account. Nevertheless a useful description of arthropod structure and function results; one that closely approximates to the basic properties of the major arthropod groups and which, with further elaborations, will become more and more a complete and detailed description of actual existing animals.

RELATIONSHIPS BETWEEN SYSTEMS

In considering the relationships between systems, the main properties of the whole animal which must be realized in every case for its very existence must be noted. These properties are:

1. A flow of food and water through the system to provide the materials and energy to build and operate the animal.
2. The flow of oxygen into and carbon dioxide out of the animal to facilitate energy production.
3. The perception of the environment and the production of an appropriate response.
4. The production of viable offspring.

Not one of these properties can be accomplished by a single system: all systems must interact for their production.

As will be apparent from the descriptions of system function already given, similar functions can be performed by systems of quite different design. A function may be performed by tissues of several different systems as a second function, such as the loss of nitrogen through the gills, gut and sometimes coelomic organs of Crustacea. Finally, functions of one system may be performed by a newly emerged system.

Any scheme to indicate the relationships between systems must recognize the need for the presence of them all and for some level of interaction under all circumstances and at all times. However, the integration between some systems may be closer than that between others and some systems may dominate others, either in the fundamental structure of the animal, or within an individual where the requirements of one system may dominate at one time and those of another at another time. Finally, interactions between systems may be largely one-way, the products of one system being necessary for the functions of another but with little reverse action occurring.

The relationships can be expressed in diagrammatic form (Fig. 6.1), which will also permit indication of the different relationships at different times and in different animals.

The boundary of the animal is indicated by a large circle within which six smaller circles representing different systems are drawn touching the boundary circle. Arrows crossing the boundary indicate the direction of flow of materials or interactions between that system and the animal's

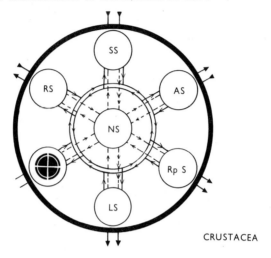

CRUSTACEA

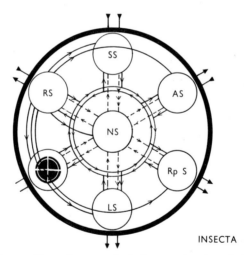

INSECTA

Fig. 6.1 Diagram illustrating a method of expressing the relationships between systems in the different arthropod groups, at different developmental stages, or at different physiological states. SS = sensory system, AS = alimentary system, RpS = reproductive system, LS = locomotory system, RS = respiratory system, NS = nervous system, ⊕ = another system. The dotted lines radiating out from the nervous system indicate flow of information to and from the system marked by arrows on the dotted lines. The arrows on the periphery indicate flow of information, or materials into and/or out of each system. The double circle between NS and the other systems represents haemocoelic circulation, the arrows to and from these

environment. The circulatory system is represented by two circles lying within the inner margins of the system circles. Arrows indicate the circulation of the haemolymph, and their opposite direction of pointing, that blood flow is haemocoelic in nature. Line-bearing arrows join the system circles to those of the circulatory system: this indicates that the systems draw from the blood in a manner which does not dictate blood circulation through systems in a fixed sequence. The arrows indicate exchange of materials between systems and the haemolymph. At the centre of the circle is a system circle labelled NS (nervous system) joined by a pair of dotted lines to all the system circles (including the circulatory system). Arrows pointing in opposite directions indicate the exchange of nervous information. The degree of exchange between systems through the circulatory system is indicated by the number of arrow heads on the line. For example, while the greater part of information from the sensory system flows into the nervous system, indicated by the presence of three arrow heads, relatively little outflow from the nervous system to sense organs is known to occur and this is shown by only one arrow head. Where no exchange through the system with the environment is found, no arrows occur: where it is unidirectional all arrows point in the same direction. Finally, the nature of the exchange product can be written and appended to the appropriate arrow.

Such a system as described would fit an adult arthropod depending upon respiration through gills or lung books and a respiratory pigment. If however, respiration were through the tracheal system, no respiratory gases would be carried to the tissues by the blood as the respiratory system would carry air directly to the other systems. This is indicated in the diagram by a series of arcs drawn just inside the boundary circle, starting from the respiratory system circle (RS) and ending at every system circle so supplied. Relationships between systems not involving circulatory or nervous systems can be indicated in this manner. Important relationships between systems through the haemocoele can be expressed by lines arising from one system passing as an arc as far as the related system and then returning. Similar lines dotted and drawn to and from the nervous system circle can express nervous connections of unusual importance. Changes in relative system size can be indicated by different diameters of the system circles. The student may like to explore this method of expressing relationships between systems by preparing diagrams to indicate the reproductive phase

circles to each system represent exchange of materials between the haemolymph and the system. For tracheate arthropods (Insecta) the respiratory system supplies air direct to the other systems, as indicated by the arrows directed from this system. In other arthropods it is conveyed via the haemolymph. For further explanation see text.

of either male or female animals. Such diagrams are useful in considering
the next step, that of expressing the total organization of arthropods.

THE BODY FORM (Fig. 6.2)

The simplest basic concept of the arthropod body form is one in which
there is a small anterior non-segmented acron bearing the eyes, followed by

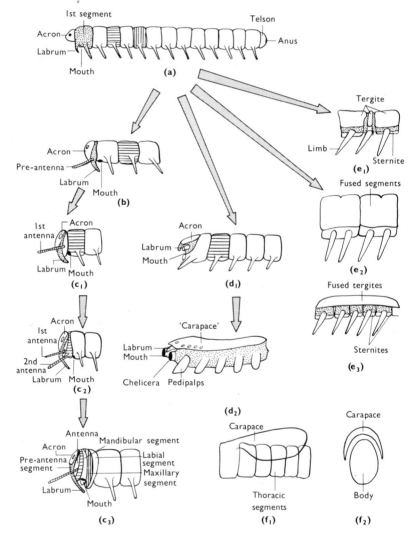

a series of uniform segments ending in a posterior non-segmental region, the telson. The mouth opens ventrally on the first segment, and the anus on the telson. In front of the mouth there is a downward projection, the labrum. There is some specialization of the appendages of the anterior segments to aid in manipulating food and sensing the environment. The coelomic organs of one of the segments are modified to form the ducts of the reproductive system. The body is elongate and approximately circular in cross-section. No living arthropod fits this concept but the Pre-Cambrian *Aysheaia* appears to have resembled it closely.

To fit this concept to present-day arthropods account must be taken of the specialization of different segments for different purposes and their association together to form body regions called tagmata, e.g. the head and trunk in Myriapoda; the head, thorax and abdomen in Insecta.

In all arthropods except in the Tardigrada, where the mouth is terminal, there is a pre-oral cavity, formed by a downward growth of the labrum and clypeus to overhang the front of the mouth. Accompanying this growth in the embryo is a movement of the anterior lateral mesoderm blocks which move forwards and grow downwards to meet in the mid-ventral line in front of the mouth, taking up a pre-oral position. In Onychophora the pre-antennal segment becomes pre-oral in position; in Chelicerata evidence favours one pre-oral segment bearing the chelicerae; in Myriapoda and Insecta there are three pre-oral segments, the second bearing the antennae; in Crustacea there are also three, the second bearing the antennules, the third the antennae. In the post-embryonic animal none of these segments are visible externally except where they bear appendages; internally their ganglia unite with the brain, and coelomic organs may persist, as in the antennary glands of Crustacea.

Some of the segments immediately following the pre-oral ones become specialized for the manipulation of food and form the lateral and posterior regions of the pre-oral cavity. In the Myriapoda, Insecta and Crustacea the appendages of the mouth-bearing segment (4th segment) form jaws, the whole limb biting at the tip in the first two groups, at the base in the third

Fig. 6.2 The body form of arthropods. (**a**) Diagram of a concept of the primitive body form of an early arthropod. (**b**) Hypothetical form with one pre-oral segment. (**c**$_1$) Hypothetical form with two pre-oral segments. (**c**$_2$) Hypothetical form with three pre-oral segments. (**c**$_3$) Segmental structure of an insect head (diagrammatic) with three pre-oral and post-oral segments. (**d**$_1$) Single pre-oral segment. (**d**)$_2$ Fusion of acron, pre-oral segment and dorsal tergites of post-oral segments to form the prosoma of arachnids (diagrammatic). Structure of the trunk. (**e**$_1$) Alteration in size of the trunk segments. (**e**$_2$) Fusion of trunk segments into pairs forming 'diplosegments'. (**e**$_3$) Fusion of tergites of adjacent trunk segments, the sternites remaining free. The carapace in Crustacea. (**f**$_1$, **f**$_2$) The carapace formed by an outgrowth of the maxillary segment : later fusion gives the typical carapace of the decapod Crustacea which differs from that of Arachnida (**d**$_2$).

group. In Onychophora the jaws are formed from the appendages of the second segment. Behind the mouth the appendages are modified to form slime papillae and have no part in the feeding mechanism of the animal. In some Myriapoda the appendages of the segment behind the jaws form a lower lip and posterior boundary to the pre-oral cavity; they tend to fuse in the mid-line. In other myriapods and in Insecta they are adapted to assist the mandibles in the manipulation of their food and it is the appendages of the second segment following the mandibles which form the lower lip of the pre-oral cavity. In Crustacea and Chelicerata fusion in the mid-line of post-oral appendages does not occur, although the appendages of the following two segments may be very deeply committed to food manipulation. In Chelicerata the limbs following the pedipalps have gnatho-bases used in food manipulation, but the limb still also has a major role in locomotion. These segments become associated together to give the most anterior tagma of the arthropod body: the head, composed of the pre-oral segments, together with the gnathocephalon, comprising those segments specialized for the manipulation of food.

In Onychophora the head is composed of the acron, pre-antennary segment, mandibular segment and the slime papillae segment. The fourth segment is the first of the uniform trunk series. There is no sharp demarcation between the last head and the first trunk segment.

In Myriapoda the head consists of three pre-oral segments together with two (mandibular and maxillary) or three (mandibulary, maxillary and labial) segments forming the gnathocephalon. The last head segment is sharply marked off from the first trunk segment and the head tagma can move relative to the rest of the body. In Insecta three pre-oral and three gnathal segments closely welded into a compact capsule compose the head. Between the head and trunk is a flexible neck region permitting considerable movement of the head relative to the trunk.

In Crustacea the head is also composed of three pre-oral and three post-oral segments. Its demarcation from the trunk is less pronounced than in myriapods and insects and a neck region is not present. In many Crustacea the maxillary segment of the head extends as a great backward fold, the carapace, lying over the dorsal and lateral parts of the following trunk segments. Primitively it is quite free from them, but in some groups it fuses with the dorsal region of these segments giving a uniformly rigid shield over the anterior body region and tending to obscure dorsally the boundary between head and trunk. Evidence suggests that the growth of this carapace has occurred independently in the several lines of crustacean evolution. This condition must be distinguished from a similar-looking dorsal plate which covers the anterior body in the Chelicerata; this is formed by the fusion of the tergites of the anterior segments and not by a backward fold from one of them.

Behind the head the remaining segments of the body plus the telson make up the trunk. In Onychophora and the Myriapoda and some phyllopod Crustacea no further sub-division into tagmata occurs. The trunk segments are uniform in the Onychophora but in the Chilopoda differentiation between them may occur. For example, in *Lithobius* the tergites may alternate in size, the sternites remaining uniform. Some of the middle trunk segments may fuse. In Diplopoda the segments fuse in pairs to form diplo-segments. These modifications affect the trunk as a whole and do not allow sub-divisions to be recognized.

In many Crustacea further sub-division of the trunk into tagmata does occur. These divisions are based on an anterior specialization of limbs for locomotion by walking, and for food collecting, the thorax; and a posterior specialization either for swimming, for brood carrying or for escape reaction, the abdomen. Where walking and food collecting are the specialized functions of the thoracic limb, the abdomen is often limbless. In the Malacostraca the thorax is of eight segments and the abdomen of seven (six in most members of the group). The appendages of the anterior thoracic segments may be modified for manipulating food into the mouth as well as for locomotion and are maxillipeds; the remaining five pairs are walking limbs and bear the gills and gonads. The abdomen bears swimmerets used in gentle paddling.

In the Insecta the trunk is formed from two tagmata called the thorax and the abdomen. The thorax is composed of the three segments following the head and bears the walking limbs. The abdomen lacks walking appendages though styli may be present; the gonads are in the abdomen and the reproductive openings are on the last segments. Posteriorly the last segment bears a pair of sensory styli, the cerci and a long medium terminal filament. The thorax and the abdomen are thus very differently composed tagmata compared with those given similar names in Crustacea.

The body form of the Trilobita tends to be that of a simple oval, flattened dorso-ventrally. In transverse section the body is expanded laterally to cover almost completely the legs. These dorso-lateral expansions, the pleura, are marked off from a central cylindrical axis by longitudinal grooves on the dorsal surface and by appendage articulations ventrally. This trilobed structure gives the group its name. There is a head tagma formed from the fusion of the more anterior segments and bearing a large labrum in the mid-ventral line, a pair of antennae arising ventrally and, on the dorsal surface, a pair of compound eyes. Behind the head are trunk segments, each bearing a pair of similar appendages; posteriorly the segments are fused together dorsally to give a terminal tagma, the pygidium. The anus is posterior and ventral.

In the Chelicerata the body tagma consist of an anterior prosoma comprising one pre-oral segment and five post-oral segments, the latter bearing the walking limbs; and a trunk, the opisthosoma, devoid of walking limbs, but often distinguishable into two further tagmata, the mesosoma and the metasoma.

Dorsally the tergites of the prosoma are fused to give a single sclerotic plate called the carapace (not homologous with that found in Crustacea). Anteriorly the carapace has a pair of median eyes and laterally one or more pairs of simple eyes. The first pair of limbs, the chelicerae, lie in front of the mouth: post-orally the prosomal limbs are set closely together and the sternal plates are reduced. In *Limulus* the carapace is extended laterally to cover the legs and anteriorly to project forward over the mouth, giving a large hood-like prosoma completely concealing the appendages from the dorsal view.

The opisthosoma lacks walking appendages. Its anterior mesosoma bears genital openings on its first segment and, in *Limulus*, plate-like appendages bearing respiratory gills on its more posterior ones. In Arachnida, lung books or tracheae occur on these segments, though often in reduced numbers. The segments of the metasoma, when present, are usually sclerotic rings much smaller in diameter though often longer than those of the mesosoma. The anus opens on the terminal telson. The metasoma is often reduced to a thin filament (whip scorpions), or absent as in spiders, solifugids and ticks. In the scorpions it is well developed and bears a terminal sting which the animal can use by striking forward over its head. Where the metasoma is reduced the mesosoma often loses its segmental sclerites and forms a uniform soft abdomen, as in most spiders. Mobility is often increased between the prosoma and opisthosoma by the development of a constriction between the two, as in spiders. In *Limulus* the opisthosomal sclerites are fused to form a single dorsal shield which also projects laterally; the anus is at the ventral tip of the shield and there is a long terminal post-anal spine formed from the telson. In other chelicerates the segments of the mesosoma and metasoma are well marked.

SYNTHESIS

The differentiation of the arthropod body into groups of segments (tagmata) (Fig. 6.3) specialized for particular purposes occurs in all arthropods. A series of tagmata once established tends to persist in the face of other structural and functional changes. It is perhaps the most basic feature in modern arthropod groups, each of which is characterized by its tagma structure. The main tagma sequences present in the Arthropoda are given; the determination of the appropriate sequence is a first step in defining an arthropod.

The different specializations of the tagmata are expressed in the presence or absence of appendages on their constituent segments and in the

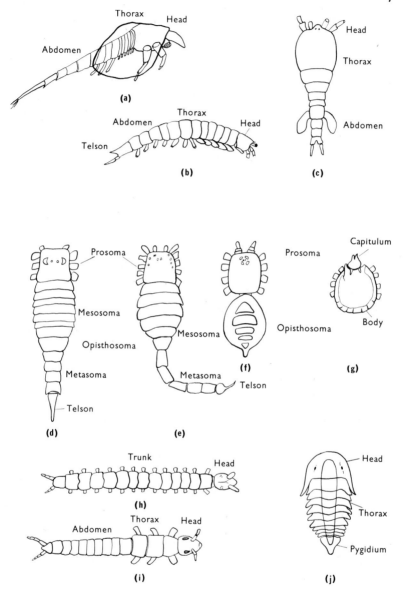

Fig. 6.3 Tagmata in the different groups of arthropods. (**a**) Crustacea (*Nebalia*). (**b**) Crustacea (*Anaspides*). (**c**) Crustacea (*Calanus*). (**d**) Merostomata (*Eurypterus*). (**e**) Arachnida (Scorpion). (**f**) Arachnida (*Liphistius*). (**g**) Tick. (**h**) Myriapoda (Centipede). (**i**) Insecta. (**j**) Trilobita.

specialization of the appendages to perform different functions. However, as appendage specialization can occur without being reflected in the body organization into tagmata, it is desirable to list separately in sequence the type and distribution of the body appendages. In the lists provided, the segment bearing the appendage is given as a number together with a letter signifying the appropriate tagma. Lack of space prevents the giving of more exhaustive lists but the student can, either by examination of specimens or attention to the principles given, prepare his own and relate them to other structural features.

Correlated with the above design features, but leaving their own imprints upon the arthropod, are a number of properties whose alternative conditions (listed below) characterize still further the organization of the animal.

Except in the Onychophora the cuticle of the animal also acts as its skeletal system, being hardened into a number of strong sclerites. These may be lost, the covering of the animal becoming a soft flexible cuticle. Complete loss is rare: almost always some part of the animal retains its skeletal features, e.g. the head capsule in insect larvae, the capitulum in ticks. Skeletal alternatives then are: primitively, its absence from the cuticle; the presence of an exoskeleton; secondary loss of an exoskeleton.

Primitive arthropod locomotion was probably walking, some degree of swimming also being possible. A commitment to swimming as a usual means of locomotion influences the appendage type and the body structure to differentiate them from those of walking animals. The ability to run applies only to terrestrial animals and this also imprints design features on the body (page 93). Flying occurs only in one arthropod group but greatly influences its structure. It is therefore important to consider whether the animal is basically a walker, swimmer, flier, runner, or a combination of several of these.

Three basic types of food manipulating systems can be distinguished in arthropods:

1. Microphagous feeding. Here a number of post-oral appendages are involved in filtering out relatively minute particles from the medium and passing them forward to the mouth, where they are pushed in by the mandibles.

2. Particulate feeding. Here individual particles are picked from the environment, or bitten off a larger mass and conveyed to the mouth. The particles may be obtained either by mandibles or by specialized post-oral appendages.

3. Liquid feeding. Here animals are caught by specialized appendages, usually chelicerae or chelae and held to the mouth; regurgitated digestive juice digests flesh externally and the juice only is sucked back into the gut.

In further evolution the equipment of microphagous and particulate feeders is often modified to permit widely different feeding of all types. These secondary modifications are imprinted upon the basic ones. Liquid feeders seem to remain liquid feeders, although much further modification occurs in order to exploit different liquid sources.

The correlation between the sensory-nervous system and the design features already mentioned needs to be considered at two levels. At the first level three types of basic design may be considered: Type I (Fig. 5.4a), where the nerve trunks are widely separated and the segmental ganglia scarcely developed. Type II where the trunks are close together and a well developed ganglion occurs in each trunk segment. Here there is only one pre-oral segment in addition to the nervous mass of the acron incorporated to form a brain. Type III (Fig. 5.4b, c, d,) is similar to Type II but three segments are pre-oral, thus giving a more complex brain.

At the second level we are concerned with differential development of the ganglia in relationship to specializations of their segments, especially their appendages. Thus a predominantly sensory appendage will bring to that ganglion a large number of inflow axons with increased size and complexity over its fellows, or the limitation of appendages to a few segments will increase the size of their ganglia and the connections between them. Consultation of the appendage list will do much to suggest nervous specialization of this type.

In many cases the nerve ganglia of a tagma will fuse to form a common mass. The concentration of appendages on the prosoma and the differentiation of some of them to sense organs allows the development of a larger nerve mass than occurs in brains and the integration of more elements to greater complexity than three pre-oral segments can achieve. Such specializations do not occur in Type I nervous systems.

The systems concerned with the flow of materials through the body and the relationships between them are also correlated with the design features already described.

The simple tubular alimentary canal designated Type I (Fig. 5.5a) provides a simple tube of epithelium occupying the centre of the cylindrical body through which material can be absorbed and lost to the animal. Transport to and from this is dominated by the circulatory system (Type II Fig. 5.5c). The relative slowness of this system is offset by the transfer of the storage capacity of the mid-gut cells to the mesodermal cells (fat body) lying in the haemocoele and forming a sheet capable of penetrating into contact with all the body tissues so that very short distances separate the storage cells from the using ones. Once a fat body is established the mid-gut can shorten without causing a corresponding increase in transport difficulties. Excretion through the gut tends to be similarly facilitated by the Malpighian tubules which penetrate throughout the haemocoele and come into contact with the user tissues. This clogging of the haemocoele reduces circulation speed still further and would not be so successful if the oxygen carriage to the tissues were not accomplished by some means other than the circulatory system, namely by the tracheal system.

Where the alimentary canal Type II (Fig. 5.5) is present together with the circulatory system Type I (Fig. 5.6a) transport is still very efficient. In these animals respiration is effected through blood pigment and ionic regulation through a single pair of coelomic organs. The necessity for transport between these prevents reduction of the distributive side of the circulatory system and over-clogging of haemocoelic space. However, increased branching of the mid-gut reduces the distance between store and user tissue. The distributive side has thus a priority over the collector side. In some animals, e.g. ticks, the respiratory side is taken over by tracheae and thus the way is open for more clogging of the haemocoele. These types of relationships should be borne in mind when selecting the alternatives from the list provided.

By taking a selection from the matrices (Chapter 11) it is possible to build up an arthropod: more details of its properties can be culled by reference to the descriptions given of systems, organs, tissues and cells so that a fairly comprehensive picture of the structure and function of the animal can be portrayed. It is not, of course, complete because certain factors have so far not been discussed or entered on the chart. In living groups not all possible combinations prepared from the chart occur and could not, due to functional inconsistencies: for example, a liquid feeder would have little use for a grinding system in its gut; gills and wings are not likely to co-exist in the same creature. It is dangerous to say that a particular combination is impossible: rather the complexity which would follow from the presence of some systems in the same body at the same time makes their occurrence less probable than other combinations.

The chart also concentrates upon the basic qualities of arthropod structure. In the evolutionary history of the groups these may be modified, as, for example, when the particulate feeding mandibulate insects adapt to fluid feeding. The mechanism bears many imprints of its origin and thus differs greatly from liquid feeding mechanisms in the Arachnida. These secondary modifications will be dealt with later and can then be written into the specification sheet. The greater patterns of arthropod organization form the groups into which they are organized and named, thus:— Onychophora, Trilobita, Myriapoda, Insecta, Crustacea, Merostomata, Arachnida, Pycnogonida.

7

The Life Patterns of Arthropods

INTRODUCTION

The life pattern comprises the ordered changes of growth, the development and the dynamic properties of the animal that occur from the moment it is formed until it dies. A life starts with the formation of the zygote. Where fertilization does not occur, the corresponding point is the moment when development commences, indicated by a rise in the oxygen consumption of the egg. Many factors can bring about the end of a life, but, as a standard for reference against which shortcomings and deviations can be compared, that commonly called old age will be taken as a proper end.

In every life the zygote is followed by embryonic developments, when the main structures of the body are laid down. Then the animal hatches and a period of post-embryonic growth and development occurs. This phase ends and the animal becomes an adult when the sexual processes start. Although changes in body form rarely occur after the commencement of the reproductive phase, growth in size may continue until either a definite body size is reached or death intervenes. It is not known whether death through old age is a programmed phenomenon in arthropods.

The life of any animal is a result of an interaction between the information contained within the zygote and the external environment in which this information is decoded and the organism grows and develops. If the information is incomplete or interpreted in the wrong sequence, development cannot proceed beyond the point at which the error occurs, and the animal dies. In many arthropod populations kept under ideal conditions some animals die without reaching maturity, death being due to genetic maladjustment.

THE DEVELOPMENT OF THE EMBRYO

The zygote, and later the embryo, are enclosed in protective membranes which also support the developing animal. Only when the yolk reserves are nearly exhausted does the animal free itself from these membranes and interact as an unprotected organism with its environment. Usually by this time the organism has reached a high degree of anatomical and functional complexity. The membranes which enclose the egg and embryo may be derived from the ovum itself or from the follicle cells which surrounded it in the ovary, or from the oviduct. Usually one of these develops a thick strong cuticle (the chorion) on its outer surface. Where it is secreted before the egg (Plate 3) is fertilized, special pores (micropyles) through which sperm can penetrate are present. They appear to be absent in decapod Crustacea where a peculiar explosive action drives the sperm nucleus through the thin cuticular membrane. Throughout embryonic development increase in size and weight of the egg frequently occur due to the absorption of water through the cuticle.

During embryonic life the unicellular zygote develops into a multicellular animal, the cells moving into positions where they develop into organs and systems characteristic of arthropod organization. The precise movements of the cells and the sequence of events in organ and system development form a unique pattern in both space and time for each arthropod species. Nevertheless, comparative studies reveal a pattern of development whose elements can be found in most arthropod embryos. An account of this pattern serves as a description of the general embryology of the group.

Development of the blastula (Fig. 7.1)

The first step in development is cell division (cleavage) which converts the zygote from a unicellular into a multicellular organism. Where small amounts of yolk are present the whole egg divides first into two cells, then into four, further divisions following until a hollow blastula is formed. Nearly all arthropod eggs are heavily laden with yolk and the cells of the blastula are large; its cavity, the blastocoele, greatly restricted. Where yolk is very abundant the bulk of the egg does not divide at cleavage, but the nuclei do. Each nucleus is surrounded by an island of cytoplasm and, as further divisions occur, the nuclei move out from the centre of the egg towards the periplasm. They enter this, and undergo further divisions until the periplasm is densely populated with nuclei; then intercellular membranes appear and thus a cellular blastula is formed. Some nuclei remain behind in the yolk. This type of cleavage is found in the Myriapoda, Insecta, and Arachnida, while total cleavage is general in Crustacea except where heavily yolked eggs occur.

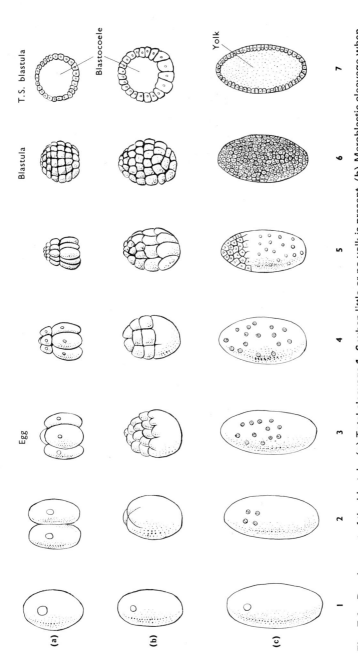

Fig. 7.1 Development of the blastula. (**a**) Total cleavage **1–6** when little or no yolk is present. (**b**) Meroblastic cleavage when considerable amounts of yolk are present. (**c**) Cleavage in centrolecithal eggs, where the nucleus divides and the daughter nuclei surrounded by islands of cytoplasm migrate to the periphery, where cell boundaries are then formed: **7** section through the blastula, the hollow space of which in (**c**) is filled with yolk.

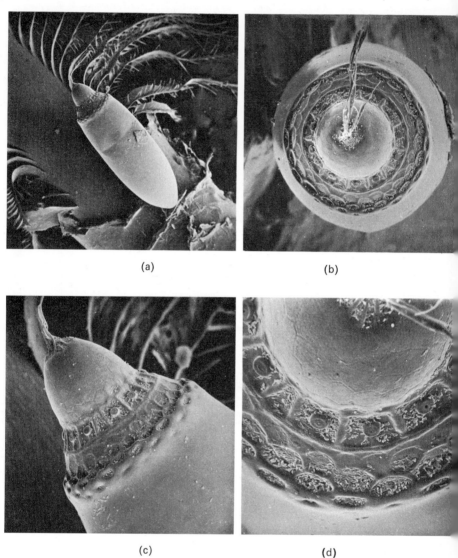

(a)   (b)

(c)   (d)

Plate 3 Stereoscan electron micrographs of the egg of an insect (louse, *Amysidea triseriata*) on the feathers of a bird (*Gallus lafayetti*) showing general appearance, and micropyles. (a) ×77, egg in situ. (b) ×260, anterior end of egg showing ring of micropyles. (c) ×310, anterior end of egg. (d) ×660, part of the ring of micropyles, one open, the others blocked. (Courtesy of the E.M. Unit & Trustees of the Brit. Mus. (Nat. Hist.), and R. Balter)

Gastrulation (Fig. 7.2)

The whole of the blastula may take part in the further formation of the embryo, but in heavily yolked eggs only a limited region (the germ disc) does so. This forms a plate of columnar cells sharply marked off from the thin cubical cells of the remainder of the blastula (extra-embryonic epithelium).

In either case, the next stage in development is the movements of cells to form the triploblastic (three-layered) type of body organization by the

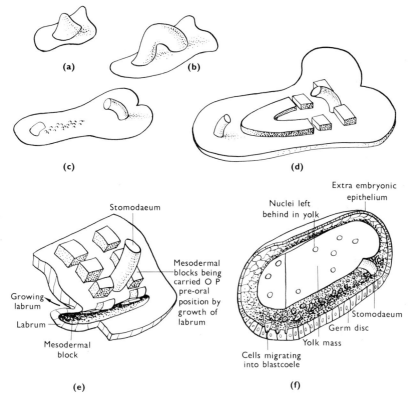

Fig. 7.2   Diagrams of phases of gastrulation in the arthropods. (**a**) Inner surface of the germ disc showing intucking of cells to form the endoderm (based on *Peripatus*). (**b**) A later stage, the embryo elongating and the intucking forming a separate mouth and anus. (**c**) Later still, fore-gut and hind-gut now separate. (**d**) Mesoderm now appearing and somites forming from its anterior end. (**e**) Later stage, diagram of anterior end showing somites being carried forward in front of the mouth by the downward growth of the labrum. (**f**) Gastrulation in an insect showing endoderm and mesoderm cells migrating in from the germ disc.

migration of cells from the surface into the region between the epithelium and the yolk. The migrating cells enclose the yolk and form the endoderm of the animal. Only in primitive arthropods is the process this simple, the formation of the endoderm (but not the migration of the cells) often being considerably delayed, and the embryo meanwhile being separated from the yolk by a provisional closure membrane. In Arachnida the endoderm is formed from the nuclei which remain behind when migration to the periplasm occurs. These cells form a layer over the yolk and under the germ disc. Following the inward migration of endoderm cells, a further migration of cells into this region occurs. These form the mesoderm, giving rise to the muscles, coelomic organs and other body tissues. Often the movements of the endoderm and mesoderm cells overlap and the nature of the migrating cell can only be determined from its fate.

The site of migration of the cells to form the endoderm and mesoderm is a restricted area on the blastula or germ disc surface. At the end of gastrulation cell movements have laid down the basis for an inner endoderm surrounding the yolk, an outer ectoderm, and an intermediate mesoderm.

Formation of the mouth and anus

The mouth and anus are formed by ingrowths of the ectoderm to form the fore- and hind-guts respectively. The inner end of the invaginations make contact with the mid-gut and eventually open into it.

In *Peripatus* a single mouth-anus invagination appears in the ectoderm just in front of the site of invagination of the endoderm and mesoderm (Fig. 7.2a–c). This opening becomes constricted into two which then move apart, the anterior forming the mouth, the posterior the anus. In other arthropods, the mouth forms first at the anterior end of the embryo, the anus later at the posterior end. A median flap, the labrum, grows from in front of the mouth and covers it anteriorly.

In further development of the arthropod embryo four major features give the phylum its special body form. These are: the development of segmentation; the development of appendages; the formation of the haemocoele; and the formation of the head.

The development of segmentation (Fig. 7.2d, e)

Segmentation first appears in the mesoderm. In *Peripatus* the mesoderm forms a U-shaped plate lying between the ectoderm and endoderm. The arms of the U lie on each side of the mouth-anus; the base is posterior. A cavity (coelom) appears simultaneously at the tip of each arm; the blocks of tissue containing the cavities separate and form a pair of hollow mesodermal blocks, the somites. Behind these, further cavities appear and become separated from the main mass. In this manner a series of pairs of somites are formed. This series is the first appearance of segmentation in the animal. In other arthropods the manner in which the somites appear varies greatly. In some the blocks separate first and the cavities appear later. Where the mouth or anus have not yet formed, the mesoderm is an oblong plate. The blocks may separate from the main mesodermal mass in anterior-posterior sequence but remain joined across the mid-ventral line, lateral separation occurring later. A series of cavities may appear in the blocks which then become divided to form simultaneously a set of somites. In any one animal the successive somites may form in different ways.

The development of the appendages

The appendages appear in the embryo as a pair of hollow outgrowths from the ectoderm of the ventro-lateral regions of each segment. They develop shortly after the mesodermal somite has been formed. Outgrowths from the mesoderm penetrate into the appendage and form its musculature. Each appendage becomes differentiated into a series of parts, the podites. Each body segment may develop a pair of appendages, but often these do not develop beyond the primary stage: regress and vanish.

The development of the haemocoele

The development of the internal organs occurs in the space between the endoderm and ectoderm, a space directly continuous with the blastocoele of the earlier embryo. This space is not obliterated, and comes to form the definite body cavity of the arthropod, which, since it also contains the blood, is called the haemocoele. The coloemic cavities formed in the earlier mesoderm become part of this cavity when the mesodermal blocks break down to form their various organs.

The development of the head (Figs. 7.2 and 6.2)

The head results from one, two or three segments developing originally behind the stomodaeum, coming to lie in front of it and contributing their mass to the pre-oral structure, together with some post-oral segments. In the early embryo, anterior to the mouth, the labrum grows ventrally and posteriorly to cover the mouth opening. The anterior mesodermal somites at this time are lying just posterior to the mouth. The growth of the labrum carries these somites forward in front of the stomodaeum. In the Onychophora the appendages of this somite form the pre-antennae. In other arthropods further segments follow the first one and come to lie anterior to the stomodaeum, maintaining their correct sequence. The three segments found in Crustacea, Myriapoda and Insecta are the maximum number moved into the pre-oral position. The antennae of these animals are derived from the second segment. The subsequent positioning of these structures near the anterior dorsal region of the head results from the growth of the germ disc to complete the animal dorsally later in embryonic development.

POST-EMBRYONIC DEVELOPMENT

Growth

During post-embryonic development the mass of living tissue and the linear dimensions of the arthropod change, relative both to time and to one another. In other animals these three factors are closely linked. In arthropods however, special circumstances uncouple them. Data about growth are derived both from observations made on individual animals throughout their growth period, and from populations of animals where each animal is observed only once. The first method is applicable where the observations do not harm the animal, e.g. linear size, weight, oxygen consumption, food intake and excretion. Such series of observations plot the actual course of changes in the animal, the only variation being that due to error of measurement. The correct analysis of such a sequence is by time series; the resolution of events in time is best studied by this method.

The second method is essential where the animal has to be killed for the observation to be made, or where handling it will introduce serious effects on subsequent observations, or where the growth of a population is to be determined. Time resolution is blurred by this method, since the variation between individuals increases the apparent duration of an event beyond that taken by any single animal.

Time series observations show that growth in weight has a component derived from the intake of food at irregular intervals, periods of fasting, loss of material due to excretion, and variation due to drinking. Imposed upon this is a trend, a tendency for the animal to change weight in a given direction over a period of time greater than that necessary for these other events to occur. The trend line may show the following patterns:

1. An increase in weight. The curve relating the values observed over the entire life of the animal to time being exponential or logistic in form (Fig. 7.3).
2. A decrease in weight, often showing an initial sharp drop followed by a shallow curve.
3. Constant weight, indicated by a horizontal line varying only by the random component.

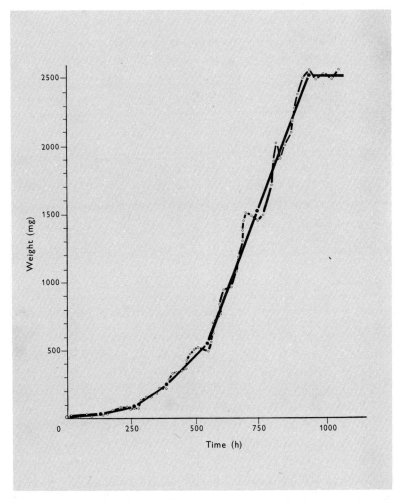

Fig. 7.3 Graph of the increase in weight of a locust during its post-embryonic growth. ○ marks each actual observation, straight lines (●—●) the trend line for each stadium.

The linear dimensions show similar trends but the random component is smaller and due to other causes.

In the newly hatched animal, the amounts of different substances present and the linear dimensions of the animal are in a definite proportion to each other. With an increase in size these proportions may remain constant (isogonic growth), or change (heterogonic or allometric growth). In isogonic growth the bigger animal is simply a large edition of the smaller one, but in heterogonic growth the shape of the animal changes (Fig. 7.4)

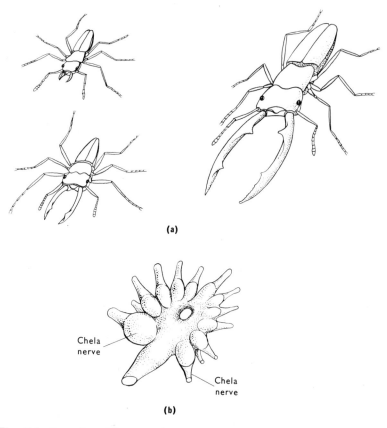

(a)

Chela
nerve

Chela
nerve

(b)

Fig. 7.4 Examples of heterogonic growth. (**a**) Three individuals of the stag beetle *Cyclommatus tarandus* showing increase of relative size of the mandible with increase in absolute body size. (**b**) Fused ventral ganglia of the male fiddler crab *Uca pugnax,* showing difference in size of the ganglia supplying the chelae which are of different size (Fig. 10.10). These differences are brought about by differential growth.

and the relative amounts of substances in it alter, so that the larger animal differs greatly from the smaller one. The relationship between two hetero-gonic variables may be expressed by the following formula

$$y = ab^x$$

I

where $y=$ the measurement of body length, $a=$ that of a part of the body, $b=$ their initial ratio, and $x=$ the calculated constant; for weight, $y=$ the weight of the animal less that of the measured part, $a=$ the weight of the part, b and x as previously.

In the larvae of the beetle *Tenebrio* sp., relative to an increase in the wet weight of the body, the factor x in the growth equation is 0·97 for total phosphorus, 0·77 for non-protein nitrogen, indicating that these substances are present in smaller relative amounts in the larger larvae. The value for fat is 1·32 and the calorific value of the tissues is 1·09, indicating that these substances are present in relatively larger amounts in large as compared with small animals.

These relationships found in all animals are also found in arthropods, but here their expression is complicated by the cyclical nature of arthropod growth. This is most marked by changes in linear size. For a period (the stadium) the dimensions of the animal are virtually unchanged, its form (instar) remaining constant. A stadium ends with the casting (ecdysis) of the old skin, the animal assuming new dimensions. The new instar may be similar to but larger than the old, or it may differ from it very greatly indeed. This method of growth is linked to the properties of the cuticle, which covers the animal, does not grow in surface area, is inelastic and forms the exoskeleton. Increase in size can only occur when the old cuticle is shed and the new one, formed under it, expanded to a larger size. These growth peculiarities can be analysed into two closely linked cycles (Fig. 7.5), each having different properties, responding to different environmental information and staggered in time relative to each other.

The primary cycle starts at apolysis, the separation of cells from the old cuticle. The cells then undergo mitotic division and produce a new epidermis of increased surface area. Since this new epidermis is contained within the old cuticle it is much folded. The form of the epidermis may be similar to the old one, or, due to heterogonic growth, it may assume new shapes and proportions. The new epidermal cells produce a new cuticle and initiate the digestion of the old endocuticle. In due course the new cuticle is completed, and the cells then rest for a period or may immediately start another growth cycle.

The secondary cycle starts at the commencement of the stadium when the animal, having been expanded to its maximum size, becomes active. The flexible intersegmental membranes become folded under the sclerotic plates. The animal may not commence to feed immediately following relaxation, but may wait for some hours to ensure that the mouthparts are sufficiently hard to function properly. Feeding then starts and the animal eats intermittently throughout most of the stadium, usually however, more in the earlier than in the latter part. Weight increases and new tissue accumulates. Towards the end of the stadium feeding ceases and the

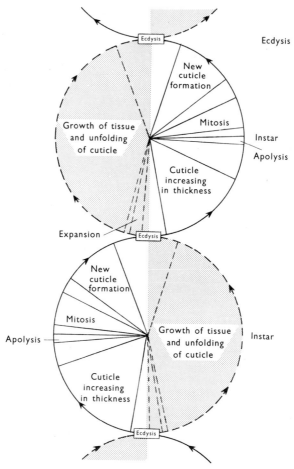

Fig. 7.5 Diagram of primary and secondary cycles in the post-embryonic growth of insects. The primary cycle solid lines indicate events in the cell structure of the epidermis. Following ecdysis the cuticle increases in thickness: apolysis, mitosis and the formation of a new cuticle and digestion of the old occur in that order. The secondary cycle is shown by dotted lines. At ecdysis the body expands in size; the cuticle hardens and then folds. During most of the instar the growth of tissue is accommodated by unfolding of the cuticle, until the next ecdysis occurs.

animal prepares for ecdysis. The creature usually seeks a sheltered place. The gut is emptied and the animal begins to swallow air, or in an aquatic one, water, thus distending the gut and beginning to build up the internal pressure of the body. This causes the old cuticle to break along predetermined lines of weakness. The animal can free itself from the old cuticle by passing through the opening thus made. The animal continues to swallow air or water and increase further in size until the whole of the new cuticle is fully expanded and this position is held until it is hardened sufficiently to retain its new dimensions. The animal then expels the air or water from its gut (Plate 4); the intersegmental membranes fold under the sclerotic plates, general activity is resumed and the cycle commences again.

Comparisons between the measurements of a sclerotic plate made at different instars of similar body form show that its size changes by a constant proportion at each ecdysis. This proportion can be calculated in two ways:

$$\text{Brook's Ratio} = \frac{\text{Final size} - \text{initial size}}{\text{initial size}} = \text{Constant} \qquad \textbf{2}$$

$$\text{Dyar's Law} = \frac{\text{Final size}}{\text{initial size}} = \text{Constant} \qquad \textbf{3}$$

Brook's Ratio is used largely in studies on Crustacea, Dyar's Law on Insecta. The 'constant' is variable, but differences amounting to twice the normal value occurring between any pair of a series would indicate that an instar of the series is missing.

The magnitude of the constant represents the amount of growth in linear size which occurs at each ecdysis. This is accompanied by an increase in volume which provides space for the tissues that will grow during the next stadium. At the moment of ecdysis this extra volume is filled by air or water contained in the gut. When the animal relaxes the gut is emptied and the cuticle folds down, largely obliterating this extra space which can be regained as required by the cuticle unfolding. In primitive arthropods this volume represents the amount of space available for growth before another ecdysis is necessary, and is governed by the extent to which folding of the cuticle, which also acts as the exoskeleton, is possible. In many arthropods special adaptations exist which allow this volume to be large, giving fewer ecdyses and more rapid growth rates, while the folding of the cuticle, incompatible with its skeletal functions, is limited. Where the gut is filled with air which is expelled at relaxation following ecdysis, some other structures such as the air sacs of the tracheal system expand and fill the vacant space, tissue growth compressing the sacs as development proceeds. In aquatic arthropods water present in the gut is often transferred to the haemocoele and then slowly excreted as the new tissues accumulate. In both the primary and secondary cycles, events, which in other animals are spread evenly in time, are concentrated into short periods.

Studies on heterogonic growth in other animals show that, as the individual grows, its proportions change gradually and throughout its life it assumes all the ratios that its growth formula specifies. This is not so in

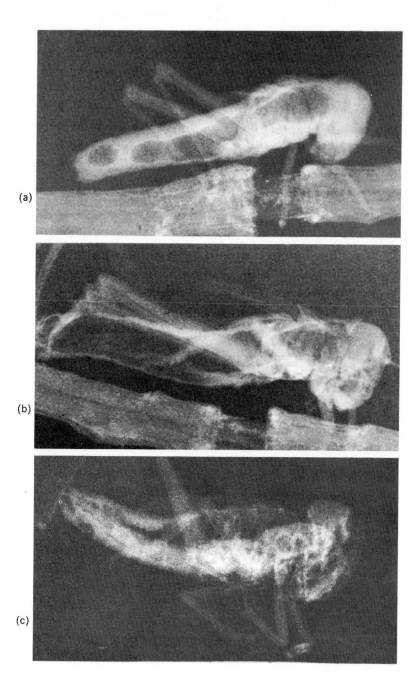

(a)

(b)

(c)

arthropods. Here there are only a finite number of steps. Measurements show that these abrupt steps give values which fall upon a normal heterogonic curve. The individual animal in arthropods may be said to jump along the growth curve, touching it at only a few points. It is important to remember the distinction between individual and population studies since, when the population is studied, individual variation gives values which, if plotted, appear to be spread along the growth curve in a manner typical of individuals of other animal groups but not of arthropods. To fail to make this distinction is to lose sight of the special features of arthropod growth. This leads to an important phenomenon. The number of instars in a species is fixed in higher arthropods but is subject to mutation, so that individuals appear having one more or one less instar than normal. Now because of the jump and the nature of heterogonic growth, individuals with an extra instar may show very different body proportions from those in normal development and individual variation may not fill the gap between them (Fig. 7.4). The differences have been so great that until these phenomena became known the animals were considered to belong to different species.

Increase in weight is nearly continuous throughout the growing period but is affected by the secondary cycle as follows:

1. At ecdysis the gut is emptied and the animal, which does not feed until the process is completed, loses weight. As the gut content forms an appreciable part, up to one third the weight of the animal, this loss is considerable.

2. In aquatic animals in which water is taken in at ecdysis, then the weight will increase at this time. Subsequent loss of water is slow throughout the following instar as it is replaced by tissue growth; hence in these cases there appears to be an increase of weight at ecdysis and a slow increase in the subsequent instar.

Metamorphosis

The formation of the new epidermis in the primary cycle takes place under the old cuticle and so is protected from the environment and free from functional necessities. It may then assume quite different proportions from those it had before mitosis commenced. The secondary cycle results in the concealment of these changes until ecdysis, when they are suddenly revealed. The differences may be small but can be very large. The abrupt appearance of the new body form at ecdysis is termed a metamorphosis.

Plate 4 X-ray photographs of 5th instar *Locusta migratoria* nymph undergoing its fifth ecdysis. (a) Air being swallowed into the gut at the beginning of ecdysis. (b) gut fully expanded, stretching the new cuticle to its fullest extent just after the skin has been cast. (c) Gut returned to normal size, cuticle folded and the extra space filled by air in the air sacs.

These abrupt changes in body form have an adaptive value. They enable successive stages in the life history to live quite different lives and thus reduce the competition that can occur between them. For example, when the aquatic dragonfly larva changes to the flying aerial adult it in no way challenges the larva for food, shelter or space. Arthropods in their post-embryonic growth pass through size ranges of several orders of magnitude. Thus the physical world of a newly hatched animal presents quite a different pattern of forces from that encountered by an animal a hundred times its bulk. Changed methods of locomotion and new skeletal proportions must by the nature of the growth of the animal be met by abrupt changes at ecdysis for the body form to keep pace with the adaptations required.

After the epidermis has detached itself from the cuticle and started to grow, its final area will be in proportion to the frequency and distribution of cell division throughout its various regions, thus producing a new folded integument which, upon expansion, will reveal the animal's new shape. The necessary changes in epidermal area are thus produced under conditions other than those which give them their final form, and the shape and size of the existing integument does not limit the change that can occur.

The new design assumed by the integument does not depend upon that of the old, but is dictated by the information contained within the cell and the pattern then read out of this information in the epidermis. Evidence for this can be obtained by observing the puffing patterns of the chromosomes and their correlation with changes in body form, by noting that mutations occur which may affect one stage in the life history but not another, and that injuries at one stage may heal and vanish only to reappear or affect the body form at another stage. These observations suggest that within the animal there is genetic information which is only activated at certain times and in connection with certain instars. When the change is small then the amount of this information is also small, but when the change is large more information is involved. This leads to the development of several distinct 'sets' of genes within a nucleus, mutually exclusive, but not necessarily assembled on the same chromosome. Activation of one set produces one kind of body form, that of another, another.

Considerable changes in shape can occur at an ecdysis and still permit the animal to remain active almost up to the moment of casting its skin and to resume activity very shortly afterwards. There is a limit to this possibility when, in addition to developing new structures, the change requires also the breakdown of muscles necessary to operate the old structures. Such changes occur in the settlement of barnacles when locomotion ceases and the animal changes from the cypris larva to the adult form. In many insects the animal may rest for a period at an ecdysis while such changes are occurring. In the endopterygote insects a long resting stage (pupa) occurs when the larva changes to the adult form but cannot resume immediate activity until some change in musculature has occurred. A further ecdysis is necessary for the pupa to become an adult as the new muscles developed in this stage cannot function until they are firmly attached to the cuticle, an event that can only occur at a moult. Once a rest period of this type has become necessary other organs become idle and can also change. Thus the gut may assume a form and function suited to an entirely different diet, sense organs may alter, and indeed in some insects an almost complete reconstruction of the internal anatomy of the animal takes place in this phase, providing us with examples of some of the most profound metamorphoses in the animal kingdom.

THE ADULT

The assumption of the adult form may occur early in life in many arthropods and is the form which exists for most of the life span. In the Insecta Pterygota the adult forms are assumed late in life, most of the life span being occupied by the larval stages. Exceptions occur, e.g. when the adult form enters diapause to survive harsh conditions, or in social insects where the queen may live for many years.

The functions of the adult in insects are mating, reproduction and, usually, dispersal. In many cases the food reserves necessary for egg production are accumulated in the larval phases and the adult animal does not feed. This is often associated with atrophy of the mouthparts and degeneration of the gut. The adult stage must be maintained for mating, dispersal and egg laying; these may be accomplished in a very short time and many insects, notably mayflies, are known for the shortness of their adult lives. In many cases the food intake in larval development is entirely used in growth and metamorphosis, the adult animal having to forage for its food.

In other arthropods the adult form is often assumed while the creature is still quite small and further growth and moulting occurs, usually with increasing intervals between ecdyses. Mating and reproduction do not occur until a certain minimum size is reached and then continue throughout the rest of the animal's life. The reproductive cycle then becomes entrained with some environmental cycle; for example, in temperate regions with the spring.

Mating, and in females oviposition, are the two most important functions to be achieved by the adult arthropod, and the external anatomy of the adult shows many adaptive features for this end.

Mating involves meeting, recognition and copulation between a male and female of the same species. In an abundant population, meeting by random chance is adequate, but often more certain meeting results from males and females seeking a common stimulus in the environment when they are physiologically ready for mating.

Once meeting has taken place recognition must occur. Often the animals possess some behaviour pattern or a specific scent, some colour or structural pattern displayed at this moment which serves to distinguish them from similar related animals. Such properties are well developed in carnivorous forms such as spiders, where the female is much larger than the male and unless a correct approach (Fig. 7.6) is made the female may not distinguish

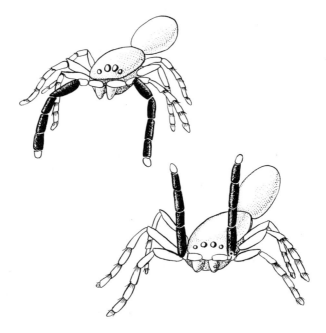

Fig. 7.6   Figures showing two positions in the courtship dance of the male Salticid spider *Corythalus*.

between the male and other edible prey. In many insects the external genitalia vary in form between the different species and the proper contact and stimuli only occur when the male and female genitalia 'fit' rather on the lock and key basis. It is perhaps for this reason that genitalia have been found so useful in taxonomic work in the separation and identification of closely related species.

Once mating has occurred oviposition follows. This may happen almost immediately, the eggs being ready for deposition and requiring only fertilization before this occurs. Frequently the eggs are not ready and the processes for egg development do not occur until after fertilization. The period between fertilization and oviposition is that necessary for egg development under the environmental conditions in which the arthropod is living. In many social insects where the queens hibernate over the winter, mating occurs in the autumn but egg development is delayed until the following spring. The eggs have to be deposited in places suitable for the survival of the young larva when it hatches. Its small size limits its ability to seek far for an appropriate habitat should it not emerge into one.

THE LIFE PATTERNS OF ARTHROPODS

The life pattern of every arthropod follows the general sequence of egg— larva—adult. This sequence may show gradual or abrupt changes at each ecdysis, the latter complicating the sequence and giving a more elaborate life pattern. The life patterns are classified according to the number of distinct forms that occur in them into monophasic, diphasic, triphasic or polyphasic life histories. This classification, which cuts across the phylogenetic structure of the group, permits a close correlation to be made between the different forms assumed and the changing environments in which the pattern is unfolding.

Monophasic life histories (Fig. 7.7)

In these no sharp discontinuity occurs in body form during growth from hatching to adult stages. In *Peripatus mosleyi* the young at birth are very like the adult. The males become sexually mature in about 9–11 months, the female in about a year. Maximum size is reached in about 3–4 years. In *Machilis* (Insecta) and most Arachnida (mites excepted), the young when they emerge from the egg closely resemble the adult although the first instar is dependent upon yolk in the gut for food. In the scorpion *Palamnaeus* there are seven instars: maturity is reached in 1–1½ years.

Over a number of moults considerable differences can arise although the actual degree of change at any one ecdysis is small. For example in the branchiopod crustacean, *Estheria syrica*, a long series of moults, at each of which additional somites and limbs are added, gradually converts the newly hatched nauplius larva into the adult form. A similar addition of somites and limbs characterizes the development of some of the millipedes (*Glomeris*) and centipedes (*Scutigera, Coleoptrata*).

Diphasic life histories (Fig. 7.8)

Here, in addition to the small changes which occur at all ecdyses, at one ecdysis there is a considerable change in body form and the animal adapted to one way of life becomes structurally suited to another. This cataclysmic ecdysis could be any one of the ecdyses that occur in the life of the animal.

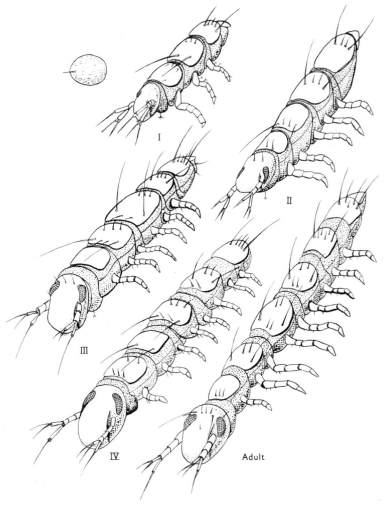

Fig. 7.7 A monophasic life history as illustrated by *Pauropus silvaticus* (Pauropoda).

It might occur at the first ecdysis as in *Limulus* where the egg hatches to a 'trilobite' larva which lacks a spine and has separate segments in the opisthosoma. At the first ecdysis the segments unite, a spine appears and the animal comes to resemble the adult. In the Crustacea the young copepod emerges from the egg as a nauplius larva. During the next six moults five additional segments and pairs of limbs are added; then, at a

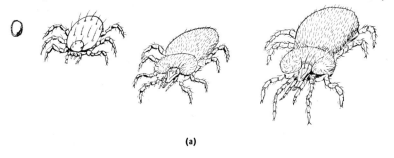

(a)

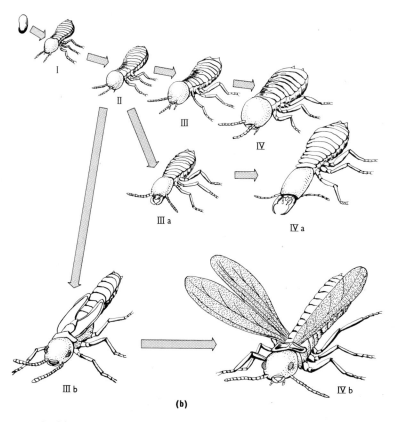

(b)

Fig. 7.8 Diphasic life histories. (**a**) *Entrombicula* sp. (Arachnida, *Acari*) showing a change of phase early in the life history. (**b**) *Amitermes atlanticus* (Insecta, Isoptera) showing major change late in the life history. I–IV development to worker, I–IVa, development to soldier, I–IVb to adult.

single ecdysis, it changes to a 'copepodid' stage which resembles the adult, lacking only the secondary sexual characters which appear in subsequent moults. In the exopterygote insects the abrupt change occurs at the last ecdysis; prior to this the larva which hatches from the egg changes its body form gradually towards the adult. In the dragonflies and mayflies where the larvae are adapted for aquatic existence, ecdyses in the larval stages only result in a slight progression towards the adult form; at the last ecdysis the aquatic form changes to a terrestrial aerial insect.

In general the cataclysmic ecdysis tends to occur either towards the beginning of the animal's life or towards the end. The former tends to be found in the Crustacea and is correlated with the larval form having dispersion as its major role; the latter in the Insecta where growth is the main function of the larval form and dispersal occurs during the adult instar.

Triphasic life histories (Fig. 7.9)

In these the arthropods have two cataclysmic ecdyses during their lives. For example, in the Cirripedia (barnacles), the egg hatches to a nauplius larva which changes to a 'cypris' larva which in turn attaches to the substratum and is transformed into the adult animal. In some Acarina (ticks),

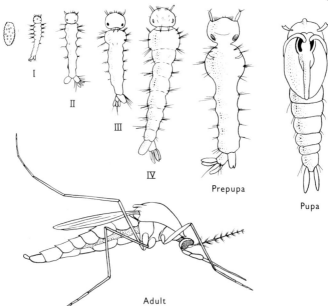

Fig. 7.9   A triphasic life history as shown by *Aedes aegypti,* vector of the yellow fever virus.

the egg hatches as a larva which moults to form a nymph which later moults to form an adult animal. In the endopterygote insects the larval form moults to a pupa. This resembles the adult in external form but lacks wings and tends to be immobile. A further moult is necessary to produce the fully functional adult insect. Finally, in many crabs the egg hatches into a zoea larva; the zoea changes into a megalopa larva which in turn moults to the adult form.

Polyphasic life histories (Fig. 7.10)

In these there are at least three catclysmic ecdyses so that at least four distinct forms occur in the life of the animal. Most decapod Crustacea (prawns, lobsters, crabs) are noted for long and complex series of changes. In the *Euphausiacea* the egg hatches to a nauplius which changes to a metanauplius, then to a protozoea and a zoea larva followed by a series of more minor changes until the adult form is reached. Amongst the endopterygote insects, in some beetles and hymenopterous parasites the egg hatches to an active larva called a 'triungulin' which searches out its host. The triungulin then changes to a soft bodied parasitic larva, which, when full grown pupates, the adult emerging from the pupa.

The maximum number of cataclysmic changes that can occur in a life history is equal to the number of ecdyses. Then, the life pattern would form a series of different forms adapted for diverse ways of life and not a progressive series leading to the adult structure. In some ticks, where the number of ecdyses is reduced to three or four, all may be cataclysmic, but usually where there are many ecdyses only a few are cataclysmic in nature.

LIFE PATTERNS AND THE ENVIRONMENT

The simplest relationship between the life pattern and the environment exists when the stages of the life pattern succeed each other without pause, egg—larva—adult—egg—larva—adult in endless succession. The rate at which these occur depends upon the supply of food and water, the temperature, the humidity and the presence of environmental signals.

The supply of materials, both in amount and quality, affects the growth and development of the animal. The needs of the different arthropods vary greatly. Near-starvation may make the animal grow more slowly, develop less quickly and live longer, but when abundant nutrition is available quick development and a correspondingly short life occur.

Temperature has a very great effect on growth, which is only possible between certain temperature limits. These vary greatly in different species and can be as low as $-2°C$ and as high as $+45°C$. Any one species, however, has a much narrower range than this, often not greater than $\pm 5°C$

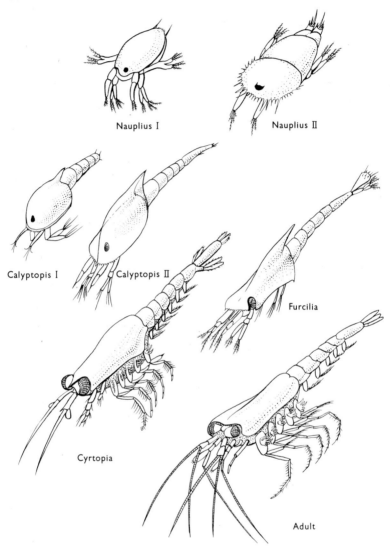

Nauplius I

Nauplius II

Calyptopis I

Calyptopis II

Furcilia

Cyrtopia

Adult

Fig. 7.10 Polyphasic life history as illustrated by a species of *Euphausiacea* (Crustacea, Malacostraca).

around its mean point. Different instars may have different ranges, those where most morphological change is occurring having the more limited range and being most sensitive to temperature changes.

Within the limits of its temperature range, any stage will grow and/or metabolize at different rates, the lowest growth rate often being two orders of time slower than the highest. The highest growth rate rarely occurs at the highest temperature at which growth is possible, since towards this limit, some temperature damage occurs. The highest growth rates are usually at 1–2° less than the upper temperature limit.

Growth rate or metabolic rate often changes regularly with temperature change within these limits (Fig. 7.11). Finding a mathematical expression to describe these curves presents many problems, partially derived from the variability of a biological system and partly from the changes not necessarily being simple variations of a common process.

The humidity of the environment has its effect through the rate of water loss from the organism, which is less in high and most in low humidities. Where stress results from high water loss and limited water intake then the growth rate is slowed.

The life pattern of the animal is correlated with the seasonal environmental changes. The physiological preparation for each environmental change must be made in advance of its occurrence. Signals in the environment which indicate the coming conditions are often utilized by the animal to initiate the necessary changes. The most commonly used signal is the photoperiod, since the ratio day/night length is correlated with climatic changes.

In writing the life patterns of various arthropods much information needs to be brought together, relating not only to the morphological states but also including statements about the speed of development, limiting conditions, steady states and alternative possibilities. Such relationships can only be stated concisely if the printed sentence is replaced by a symbol. At this stage we will introduce some short statements which can be elaborated as further information and synthesis of arthropod biology is achieved. Such propositions are made with close attention to Boolean algebra which seems well suited for this purpose. (See Bibliography. Hohn, 1960.)

First one must have symbols to represent the major morphological changes in the development of the animal: z can stand for zygote, meaning the fertilized egg of the animal; e for embryonic development; l for larva, meaning the post-embryonic to pre-adult stage of the life cycle; a for adult, meaning the body form the animal has during the reproductive cycles of the life history. The life pattern is terminated by death, indicated by d written at the end of the symbols. Thus the stages in the life pattern of an arthropod could be indicated thus:

$$z \; e \; l \; a \; d \hspace{6cm} \text{1}$$

the symbols written in sequence without punctuation indicating that the events they represent follow one another in the order stated. Where it is

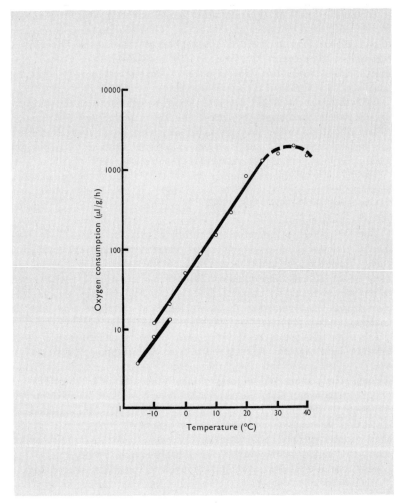

Fig. 7.11 Graph showing the relationship between oxygen consumption (indicator of metabolism) and temperature of *Anagusta kuhniella* (Insecta, Lepidoptera). Upper curve for mature larvae just before spinning the cocoons, lower curve for prepupae in cocoons. (By courtesy of Salt, 1958)

necessary to recognize more stages in the life history extra terms can be introduced, as for example, in endopterygote insects where, if p stands for pupa, the symbolic representation would be

$$z \; e \; l \; p \; a \; d$$ 2

If more than one larval stage needs to be recognized terms can be introduced thus:

$$z\ e\ n\ m\ c\ f\ o\ a\ d \qquad\qquad 3$$

for the Euphausiacea where $n=$ nauplius, $m=$ metanauplius, $c=$ calyptopsis larva, $f=$ furcilia larva, $o=$ cyrtopia larva and $a=$ adult.

Where names have not been given to the larval forms a sequence of larval stages could be designated:

$$l_1\ l_2\ l_3\ l_4\ l_5\ l_6\ l_7\ l_8 \text{ etc.}$$

Each of the stages so named has properties which must be taken into account: these can be indicated by a symbol which can be written as subscript to the appropriate stage. Since all these properties must be present they are enclosed in a bracket and as they are all of equal importance the symbols are separated by a full stop. This system of notation says all must occur; all are of equal value to the animal. These properties and their symbols are: g for growth, df for differentiation, r for reproduction, and dp for diapause.

Thus a larval stage may show growth and differentiation and would be symbolized thus:

$$l_{(g.df)}$$

or growth only

$$l_g$$

or be a resting diapause stage

$$l_{dp}$$

A sequence of larval stages showing these events could be symbolized thus:

$$l_{1_{(g.df)}}\ l_{2_{(g)}}\ l_{3_{(dp)}}\ p\ a\ d$$

An entire life history of an endopterygote could be written thus:

$$z\ e_{(df)}\ l_{(g.df)}\ p_{(df\ dp\ df)}\ a_{(g.r)}\ d$$

where $l_{(g.df)}$ indicates that in larval development growth and differentiation occur together, but in the pupal stage the absence of stops in the term $(df\ dp\ df)$ indicates that these events, differentiation, diapause and differentiation must occur in the sequence.

The rates of development and tolerances of physical conditions for each stage differ; they may be symbolically represented thus: $T=$ temperature in degrees centigrade, $R.H.$ relative humidity and $P=$ photoperiod, the number of hours of light and dark in a day to which the stage is subjected. The limits of temperature are indicated by an integral sign with the figures written the upper limit at the top, the lower at the bottom of the sign, and the photoperiod by the number of hours light/number of hours dark. Thus a larval period may be written

$$l_{(g.df)}(T\int_{10}^{30}R.H.\int_{20}^{90}P16/8)$$

indicating that larval growth and differentiation normally take place within a temperature range $10-30°C$, a relative humidity $20-90\%$, a photoperiod of 16 hours daylight and 8 hours darkness. If necessary the shape of the developmental curves for temperature can be written following the integral. For a pupa with diapause the diapause conditions can be written following the symbol in subscript thus:

$$P_{(df}T\int_{10}^{20}dpT\int_{-\frac{6}{4}df}^{}T\int_{10}^{20})$$

The formula for a complete life history of the tropical bug *Dysdercus intermedius*, a serious pest on cotton in Africa, can be written thus:

$$\approx e_{(T\int_{17.5}^{32.5}D\int_{16.2}^{4.1})}\ l_{1_{(g)}}(T\int_{17.5}^{35}D\int_{9.4}^{2.9})\ l_{2_{(g)}}(T\int_{17.5}^{35}D\int_{18.8}^{3.2})$$

$$l_{3_{(g)}}(T\int_{17.5}^{35}D\int_{12.7}^{4.5})l_{4_{(g)}}(T\int_{17.5}^{35}D\int_{13.5}^{5.6})l_{5_{(g)}}(T\int_{20}^{32.5}D\int_{16}^{6.8})$$

$$a_{(g,r)}(T\int_{17.5}^{32.5}D\int_{19.6}^{7.5}Pr\int_{8.2}^{2.1}PrO\int_{19.9}^{7.4}I.O\int_{8}^{5.5})d$$

D = duration of instar in days.
Pr   pre-copulatory period in days.
PrO   pre-oviposition period in days.
I.O.   inter-oviposition period in days.

8

The Dynamics of Arthropod Organization

INTRODUCTION

The dynamics of arthropod organization deals with the ability of the animal to perform the various functions that its qualitative properties have made possible, to co-ordinate these activities, and to relate them to the environment.

At all stages the living arthropod is exchanging energy and materials with its environment. Energy in the form of heat flows across the surface of the animal, oxygen flows into it, and carbon dioxide out of it, throughout its life. Water and food are taken in, and water and waste matter are excreted intermittently although in some circumstances the flow of water through the animal is continuous. In thermodynamic terms the arthropod is an open system, one in which matter and energy are freely exchanged with the environment.

Arising out of the properties of such a system an arthropod possesses the following features on which its dynamic functions are based:

1. The properties of the system are set by its 'constants' and not by the nature and amount of material which flows through it. The constants of the system stem from the information encoded within the zygote and thus depend upon the genetic and cytoplasmic features therein contained. These constitute the programme of the animal.
2. The materials flowing through the system supply both the energy and the substances which form it.
3. The flow of materials is irreversible due to the presence of 'valves' along the flow pathway.

4. Under these conditions flow along the pathway would normally be at its maximum capacity and input and output not related to each other or to the animal's requirements. These shortcomings are avoided by the presence of a link between input and output which is not part of the original flow path, but feeds back information about the output and adjusts the flow rate to the animal's requirements. If the action of the feedback information results in acceleration it is positive, if in deceleration it is negative information feedback. Acceleration is frequently achieved by reduction in the amount of negative feedback information. The feedback links may be simple and direct, or very complex and circuitous, when part of the chain may be through the environment outside the organism itself.

5. In either case to change the rate of flow takes time. The shorter the feedback path, the more quickly information can be sent through it and the more accurately input will match output requirements. This time lag produces oscillations in output, whose amplitude is small when simple quick-acting feedback links are present, and great when slow, long paths are involved.

The linking together of all the processes that occur in the arthropod produces an efficiently functioning organism capable of carrying out its own programme and responding adequately to its environment. It is presumed that of all the numerous conditions that could be present in an animal, for a given set of conditions there would be a fixed set of responses 'best' fitted to the situation. When these are realized the animal is then in an optimal state. Normally the animal is growing and developing in a changing environment. The output of all systems does therefore vary, either to meet present conditions or in anticipation of future requirements.

In time each of the numerous dynamic functions of the body follows a pattern that can be analysed into two components: that dictated by the varying demands made upon it by changing external conditions and the service requirements of other functions, and that due to the programmed changes demanded by the constants of the system. In general, the former have periods of minutes, hours, and occasionally days; the latter, days, weeks or months. The programmed change may be identified with the trend line calculated for a set of observations by time series analysis providing the duration of the observed period is long enough. The trend line may be one of several forms descriptive of the programmed change of the function which are:

1. A straight line, the slope of which indicates the rate of increase or decrease of the function.

2. A curve, indicating either an increase or a decrease of the value of the function in time.

3. Cyclical; that is, the trend shows a series of more or less regular oscillations.

Where mechanisms exist to return the function value to the programmed trend when, for whatever reason (environmental or service requirements), it has deviated from it, the system can be said to be a stable one. Cyclical trends are found for many functions of the arthropod body ranging from oscillations in intracellular chemical reactions to rhythmical behaviour patterns exhibited by the whole animal. All these processes have two further important dynamic features in common:

1. Stimuli affecting the rate of the function will have different effects depending upon the phase of the oscillation at which they are applied. A stimulus leading to a decrease in the amount of material subjected to a fluctuating trend will hasten the downward path and appear to have a considerable depressive effect, but when applied to the ascending path may slow it down but not stop it and thus appear to be without effect, unless of course the trend and the phase at interference are known.

2. The period of the cycle is very different for different functions and may be minutes, hours, days or weeks. A special class of oscillations is constituted by those with a period of approximately 24 hours and is known as circadian rhythms. It is often of adaptive value to the animal if they can be matched with the day-night rhythm of the environment. Coincidence may produce and maintain such a match, but usually signals from the environment entrain the rhythms together. Such signals may have to occur for each cycle or only once in every few cycles.

THE OPERATION OF THE ARTHROPOD BODY

The important properties in the dynamics of arthropod function are those which initiate a process, control the rate at which it operates and eventually terminate it. Information to set up the machinery by which a process, whether at cellular, tissue, organ or higher organizational level, can operate, stems from the programme encoded in the zygote. The previous chapters have contained an account of the kinds and functions of the machinery found in arthropods and the sequence in which they appear. Once a structure and function have been established its operation is continuous, but it may have 'nil' output to the rest of the organism. Continuous operation is a consequence of feedback mechanisms and the 'nil' output is achieved by (a) the output being absorbed completely in the feedback link, or (b) the output being below the threshold of action, or (c) the output being in a 'nonsense' form. The maximum rate at which a process can proceed is limited by the capacity of the machinery concerned in its operation and conditions set by the design of the animal. Between the limits of nil and maximum output is the operational range. Arthropods do not appear to have any exceptional capacity compared with other invertebrates in this respect except for oxygen consumption which, in flying insects, may be ten to one hundred times the resting rate. Trained human athletes have a maximum of about six times their normal rate.

The detailed mechanisms by which the rate of a process can be changed are numerous, but the following ways should be borne in mind:

1. The capacity of the machinery which carries out a process can be varied by increasing or decreasing its amount when changes in rate are accompanied by the manufacture or destruction of materials composing the system.

2. The capacity of the machinery remains constant but the idle part of it assumes an inactive form. Responses to requirements are accomplished much more quickly by this relationship than by the former one.

3. The capacity of the system remains the same but the rate is changed by alterations in the time taken for events to occur. For example, chemical reactions may be slowed by dilution (heat-adapted organisms), or hastened when water is withdrawn to give a higher concentration of reactants (cold-adapted organisms).

4. The ability of a link to convey a message may change, for example changes in the permeability of membranes.

The operational characteristics of the animal at any one moment are related to each other and to the environmental requirements by control and regulatory mechanisms, which in their turn develop by interpretation of the programme of the animal. These mechanisms form two classes, chemical regulation and nervous regulation, which themselves interact together and guide the animal through its life.

Chemical regulation

The cell is a highly organized unit within which the ebb and flow of chemical interaction is controlled and regulated by many complex intracellular regulatory mechanisms. It is upon this substratum that further chemical regulatory mechanisms operate to integrate supracellular activities; this they do by replacement of already existing mechanisms, by additions to them or by creation of new control sites.

Action between the normal cells of different tissues whereby the chemical product of one cell influences another has been clearly demonstrated in insects. In the moth *Ephestia kuehniella* a mutant form occurs in which the black pigment of the larval epidermis, testis and ganglia is missing and that of the compound eyes replaced by a red one. Transplantation of the testis from the normal male into the mutant male at a sensitive period prior to pupation leads to the development of the black pigment in the organs from which it was missing, and the red eye pigment is changed to black. Implantation of other organs from the normal moth has the same effect as does haemolymph transfusion from the flies *Drosophila melanogaster* and *Calliphora erythrocephala*.

The source of the substances bringing about these effects are the genes, and their products are the gene hormones. These observations show that the genetic information within the organisms, in spite of being located in numerous separate nuclei, can interact, regulate their output to body requirements and thus overcome errors that may arise during their frequent replication.

Chemical interaction at the cellular level serves to co-ordinate the cellular activities leading to the formation of the fully developed embryo. In the post-embryonic stages and in the adult animal, while these interactions still continue, other more sophisticated regulatory mechanisms are present. They reach their greatest development in the endocrine systems of Insecta and decapod Crustacea. Much is known about endocrine function in these animals but only fragmentary knowledge is available from the other groups.

The basic cell type on which the endocrine system of arthropods is founded is the neurosecretory cell (Chapter 2). These cells are distributed throughout the brain and ganglia of the ventral nerve cord in all arthropods. Typically they occur in groups, although single cells, bisymmetrically arranged, occur in the ventral ganglia. In Xiphosura the number of neurosecretory cells increases as the animal grows, but in other arthropods the number appears to remain approximately constant throughout life.

The neurosecretory substance produced in the cell body is conducted through the axon to the terminal arborizations where it is released. It may be conducted to and released at the site of action, particularly in small arthropods. Usually it is conveyed to its site of action through the circulatory system. The axons of the neurosecretory cells of the brain converge and form a neurohaemal organ where material is released from their terminal arborizations into the blood. This neurohaemal organ is situated behind the brain (Myriapoda (part), Insecta, and Arachnida); beneath the brain (Chilopoda); or in association with the primary optic centres (Crustacea). In Insecta it is called the corpora cardiaca and in the Crustacea the sinus gland.

In many arthropods the neurosecretory cells, their axon pathways and their neurohaemal organs form the only known endocrine system present. Evidence of its function is largely confined to correlations between the visible state of the neurosecretory cell and the general condition of the animal. The amount of neurosecretory material in the system tends to be sparse in immature forms and more abundant in adults, especially at the onset of sexual maturity. In temperate climates it often varies with the

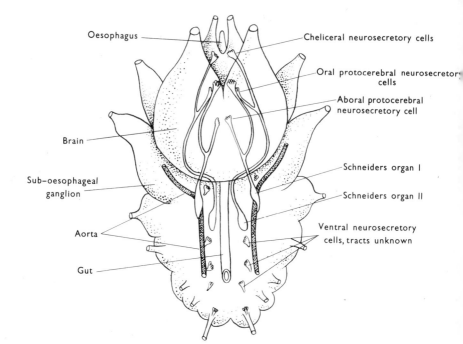

Oesophagus

Cheliceral neurosecretory cells

Oral protocerebral neurosecretory
cells

Aboral protocerebral
neurosecretory cell

Brain

Schneiders organ I

Sub-oesophageal
ganglion

Schneiders organ II

Aorta

Ventral neurosecretory
cells, tracts unknown

Gut

Fig. 8.1 The endocrine system of a spider (Arachnida). The axon tracts of the neurosecretory cell groups in the posterior part of the sub-oesophageal ganglion have not been traced.

season. In Copepoda it is scarce in winter animals but abundant in summer ones. In Diplopoda some cells (Type A) are packed full of neurosecretion in winter, others (Type B) in summer, and yet others (Type C) in the spring and again later in the summer. These variations may be due to changes in the rate of neurosecretory production. Usually it reflects differences in release rate. A cell with little of the product could be one in which no synthesis is occurring, or one in which active synthesis is proceeding but material is released as quickly as it is produced.

In Insecta other endocrine organs have evolved and become integrated with the neurosecretory system (Fig. 8.2). The corpora cardiaca is a neurohaemal organ formed in association with the terminations of neurosecretory cells whose cell bodies lie in the protocerebrum. Two groups of medial neurosecretory cells give rise to axon pathways which pass back through the brain, cross over and emerge from the posterior surface of the brain. They form the nervi corpori cardiaci I, conducting material to the contra-lateral side of the corpora cardiaca where it is stored until released

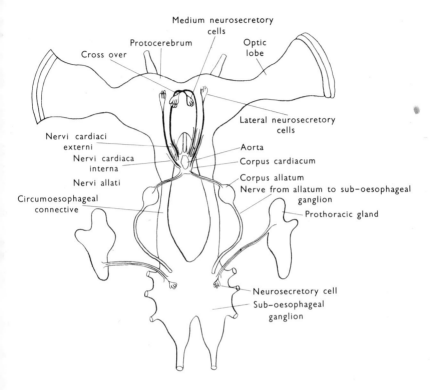

Fig. 8.2 The endocrine system of an insect, based principally on that described for primitive winged insects: in the majority of insects there is no nervous connection between the prothoracic gland and the sub-oesophageal ganglion.

into the aorta which passes through this gland. Two further groups, the lateral neurosecretory cells, also give rise to axons which pass back through the brain without crossing over and emerge as the nervi corpori cardiaci II going to the ipsi-lateral side of the gland. Besides neurosecretory axons, nerve fibres also pass from the brain to the gland via these tracts. The corpora cardiaca also has its own intrinsic endocrine cells; both neurosecretory cells and large chromophilic cells of endocrine function are present.

Some of the neurosecretory axons do not terminate in the corpora cardiaca but continue via the nervi allati to a pair of endocrine organs, the corpora allata, situated in the haemolymph one on each side of the gut. The histology of this organ has already been described (Chapter 4). The neurosecretory material is not stored in the corpora allata which produce their own hormones but it is necessary for their proper functioning, the gland atrophying in its absence. Neurosecretory material may reach the allata either through the nervi allati or from the haemolymph. Nerve fibres also pass to the glands through this nerve.

The prothoracic or ventral glands form the third endocrine organ commonly present in insects. In detail these are diverse in general form and distribution but they are frequently found in the ventral head or neck region as masses of cells, sometimes scarcely joined together, sometimes forming definite epithelial sheets. In some primitive insects they have a nerve supply arising from the ventral nerve cord but in higher forms no nervous connections are present.

The regulation of body functions by the endocrine glands commences once their anatomical organization has become established in the late embryo and it continues throughout the life of the arthropod. It is achieved through changes in the rate of release of hormones from the glands leading to different patterns of hormone concentration in the haemolymph. The cell tissues and organs respond to these hormonal patterns differently according to their nature.

The sequence of morphological changes which occurs in the post-embryonic growth of the arthropod is commanded through the actions of the endocrine glands, and in the adult they regulate the reproductive ability of the animal. At the beginning of the first instar the activation hormone stemming from the neurosecretory cells of the brain is released into the blood stream from the corpora cardiaca. The hormone is conveyed to the prothoracic gland which responds by its cells undergoing mitotic activity. This is followed by the production of the moulting hormone (ecdysone). When the prothoracic gland has been stimulated by the presence of the activation hormone for some time it is able to continue growth and hormone production independently. Any cessation of the stimulus before this period leads to the prothoracic gland returning to an inactive state. During this period of prothoracic gland activation the corpora allata also secretes its product, the juvenile hormone. The epidermal cells respond to this hormonal milieu by apolysis, followed by mitotic division and the events of the moulting cycle. The instar following the ecdysis which terminates the moulting cycle is a larger larval form.

Analysis shows that the moulting hormone, if acting by itself, would cause the next instar to resemble the adult very closely. The juvenile hormone is necessary to make the cells respond with reference to the juvenile information contained in the genes rather than to that of the adult.

This pattern of endocrine action is repeated and initiates and maintains each growth and moulting cycle during the post-embryonic development of the animal, except the last one which leads to the adult instar. During growth there are changes in the relative sizes of the body and the endocrine system, and of the endocrine glands relative to each other. The effect of this is most noticeable for the corpora allata. The concentration of the juvenile hormone relative to the others decreases as growth progresses so that activation of juvenile information in the epidermal cells becomes relatively less and some progress to adult form does occur at each ecdysis. At the commencement of the moulting cycle leading to the adult instar the corpora allata do not develop as in previous stadia and the moulting hormone acts upon the epidermal cells in the absence of the juvenile hormone. This leads to the information for adult structure being brought into action and an adult is formed which appears at the following ecdysis.

In pterygote insects, following the adult ecdysis the prothoracic gland degenerates, the hormonal milieu of the juvenile animal being necessary for its maintenance.

Early in the adult instar the corpora allata again become functional and the hormonal system of the adult animal consists of the neurosecretory and the corpora allata hormones. Both are concerned in the growth and functioning of the reproductive system. In the absence of the prothoracic gland and the moulting hormone further moulting cycles do not occur.

Experiments involving the transplantation of active glands into animals at the stage of development where the gland is inactive make good the hormonal deficiency and alter the normal pattern of growth and development. Implantation of active corpora allata into the last larval stage will supply juvenile hormone and give rise to a large larval form instead of an adult at the following ecdysis. Similarly implantation of the prothoracic gland into the adult supplies the missing moulting hormone and causes the adult to undergo a moulting cycle. Since juvenile hormone is present juvenile information is activated and the animal takes on a juvenile form at the next ecdysis. These processes cannot be pushed too far. Normally only two such moults can be forced on any one animal. Enormous juvenile forms cannot be produced nor can the animal be made to regress.

In the Endopterygota, where the change from larval to adult form is through a pupal instar, the information for pupal formation is called forth from the cells by a pattern of activation, moulting and juvenile hormones in which the juvenile hormone is relatively less abundant than the other two. The prothoracic gland persists during the pupal moult and the corpora allata are inactive over the change from pupa to adult, the prothoracic gland degenerating early in the adult instar. In Diptera patterns of chromosome puffing under the influence of different hormonal patterns mimic the normal changes seen in growth from larva to pupa and pupa to adult. The moulting hormone can be observed to produce activation of a gene $(1-18-C)$ followed by a definite sequence of puffs. Whether this is due to direct action of the hormone on the gene or on alteration in nuclear membrane permeability is uncertain.

Deficiencies in the hormonal complement may also lead to the arrest of growth and development when the animal assumes a low energy state known as diapause which is extremely resistant to adverse climatic conditions. Diapause may result from the absence of the activation hormone and the consequent absence of the moulting hormone. Once this has happened a long period has to pass before the system can recommence to operate. This period has often to be passed under environmental conditions which are very different from those under which the system is active. In *Cecropia* pupae, once inactivation has set in, only a prolonged chilling of the brain can allow the system to re-operate. Implantation of chilled brains into recent diapause animals will break diapause, as will transplantation of activated prothoracic glands. In the adult stage diapause usually affects only the reproductive organs and results from deficiencies of both the activation hormone and the juvenile hormone.

Diapause in the embryo is usually determined by the mother. In silkworms the eggs are affected by a special neurosecretory material from the sub-oesophageal ganglion of the female parent; in the absence of this material the eggs do not diapause.

The metabolic rate of the animal is also influenced by hormones secreted from the endocrine glands of insects, as follows:

1. The endocrine command to grow or change will lead to increased metabolic activity. The increase that follows can result from the design and linkage of the system together and not from numerous action sites of the hormone. In normal females of the bug *Pyrrhocoris* implantation of active corpora allata results in increased ovarian development and increased oxygen consumption. No increase in oxygen consumption occurs in the absence of the ovaries.

2. The metabolic systems may be directly stimulated by hormones. The rate of protein metabolism in the cells is influenced by neurosecretory material, by hormones from the corpora allata and by the moulting hormone. Fat metabolism is influenced by the corpora allata and carbohydrate metabolism by all three as well as by the intrinsic hormones of the corpora cardiaca.

3. Hormones may facilitate the rate of flow of material throughout the body. Neurosecretory material influences the excretion of material from the Malpighian tubules. The myotropic hormones from neurosecretion, corpora allata and corpora cardiaca increase the movement of heart, gut and Malpighian tubules. The permeability of membranes may well be influenced by the corpora cardiaca hormones.

In decapod Crustacea the neurosecretory cells occur in many groups spread throughout the brain and ventral ganglia. Axons from these groups converge to form neurohaemal organs, the sinus glands, the postcommisural organs and the pericardial glands (Fig. 8.3). The sinus glands which lie on the optic lobes of the brain receive fibres from many groups of cells, some lying close beneath the neurilemma of the optic stalk, some situated deeper in the brain or further back in the ventral nerve cord. The product of a cell group often shows distinctive staining characteristics and these tinctorial peculiarities are retained by the material when it is stored in the sinus glands.

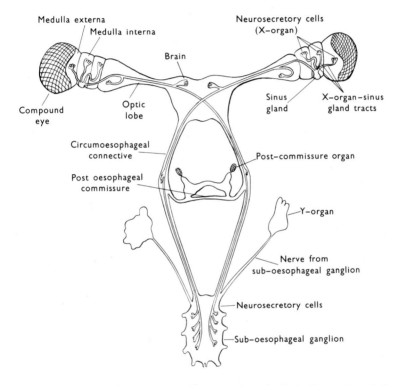

Fig. 8.3 The endocrine system in Crustacea, based principally on that of the decapod *Leander*.

The postcommissural neurohaemal organs (Fig. 8.3) occur on the commissure lying just behind the oesophagus. They receive axons arising from neurosecretory cells situated in the tritocerebral region of the brain.

The pericardial neurohaemal organs lie in the pericardial cavity. Freely bathed in haemolymph, they are well placed to release hormones into the blood as it enters the heart. Intrinsic neurosecretory cells are present in the organ and axons pass to it from two groups of cell bodies situated in the ventral ganglionic mass.

Two other endocrine glands are associated with the neurosecretory system in Crustacea. One is the ventral gland or Y-organ (Fig. 8.3). In crabs it lies in the vicinity of the mandibular muscles; histologically it is very similar to the prothoracic gland of insects. The other gland is the androgenic gland which usually consists of a strand of cells lying along the vas deferens near its termination in male Crustacea. In some Crustacea the ovary secretes a hormone which controls the secondary sexual characters of the female.

As in insects the hormonal system serves to guide the animal through its life, integrating events within the animal and between the animal and its environment. We have no knowledge of the functioning of the system in the early stages of post-embryonic growth and knowledge of its effects on growth and moulting is confined to animals which have already assumed the adult form but are not necessarily sexually mature. Following ecdysis the hormones of the sinus gland inhibit the growth of and the release of the moulting hormones from the Y-organ. Removal of the Y-organ stops the moulting cycle completely. Its moulting hormone appears to be very similar to that of insects, since ecdysone can replace the hormone in crustacean systems and vice versa. There is no gland corresponding to the insectan corpora allata in Crustacea, but it should be remembered that studies have been confined to animals of adult form. In the adult sexually mature animal maturation of germ cells in the gonads is inhibited by hormones from the sinus gland; in their absence maturation occurs. The ovaries may increase to a size thirteen times that found in the normal animal (*Palaemon serratus*).

The androgenic gland determines the primary and secondary sexual characters of the male. Implantation of this gland into the young female will cause the ovary to produce sperm and its secondary sexual characters to become masculinized. This latter is a direct effect of the androgenic hormone since it occurs also after implantation in castrated females.

Termination of growth in Crustacea may or may not accompany the assumption of the adult form and/or sexual maturity, or the animal may grow throughout its life. Where adult size is definite and growth comes to an end, this is accomplished either by degeneration of the Y-organ, as occurs in the crab, *Maia*, or by a continuous release of neurosecretory inhibitor from the sinus gland, as in *Carcinus*. Diapause is not known to occur in Crustacea.

Other body actions are also controlled by hormones released from the endocrine glands. Hormones released from the sinus glands depress oxygen consumption; the removal of these glands leads to increased oxygen utilization. Changes in blood sugar level also result from sinus gland hormones;

it is lowered in *Panulirus*, but not in many other decapods, following removal of the eyestalks. Effects on water uptake and protein metabolism have also been attributed to sinus gland hormones.

Crustacea have an elaborate system of chromatophores, and pigment present in them and in the distal pigment cells of the eye can be aggregated or dispersed bringing about colour changes in the animal. The movement of these pigments is under hormonal control.

Concentration of pigment in the distal pigment cells of the eye is brought about by a hormone, the dark-adapting hormone, so named because this is a characteristic condition of the dark-adapted animal where the hormone is in high concentration in its blood. Another separate substance, the light-adapting hormone, brings about dispersion of the eye pigments. The position of the pigment at any one time is determined by a balance in concentration between these two antagonistic hormones.

The chromatophores of the epidermis are controlled by hormones from the eyestalk and the postcommissural organs. Pigment-concentrating and pigment-dispersal substances have been separated from both organs and each pigment type may be affected separately, independently of the other, and more than one substance producing the same effect may be present. Pigment control is also influenced by the direct action of light upon the chromatophores.

Nervous Regulation

In addition to chemical regulation, integration of function within the body and between it and the environment is accomplished through information conveyed by the nervous system. Basically, regulation by the nervous system affects the functions of the animal in much the same way as does chemical regulation. Indeed the final transfer from the nervous system to the controlled function is through a chemical transmitter and its effect is to start, stop, or change the rate of a function. Nervous regulation achieves greater selectivity, quicker response times, and permits the development of more complicated patterns of instruction than are found in chemical regulatory systems. It is not a substitute for chemical regulation: both processes often affect the same function.

The nervous system functions continually throughout the life of the animal although parts of it may be electrically silent for considerable periods. Its capabilities are established by its micro- and macro-anatomy and the fixed properties of its neurons. Nevertheless its dynamic functioning, the varying inflow of sensory information, and the changing output of motor instruction are the features which help to guide the animal throughout its life. The relationship between this inflow and outflow comes from the integrative properties of the system, properties probably established from information contained in the zygote and perhaps subjected to controlled programme changes throughout the life of the arthropod.

The amount of sensory inflow into the animal is enormous and continuous throughout the animal's life. It will tend to be large in highly diverse and variable environments and minimal in constant and uniform ones. Because of adaptive phenomena found in many sense organs it is probably maximal during change from one set of conditions to another. In some cases the sense organs receive nerves from the central nervous system which modify their properties and perhaps silence their output.

The output from the nervous system is a pattern of nerve messages since the accomplishment, by the whole animal, of any act under nervous control requires action at more than one site. The output is patterned in various ways:

1. A pattern is created in the central nervous system and sent out over normally silent neurons.

2. The pattern is created by the synchronization of an output always present, but meaningless until so ordered.

3. The pattern is always present, regulation coming from changes in its rate.

Reflex activity, a given sensory input producing a given motor output, dominates the integrative activity of the arthropod nervous system. The simplest anatomical arrangement to achieve this is a single neuron. In the hearts of Crustacea the terminal arborizations from a part of the axon may act as receptors, the axon as the link and other terminal arborizations as the motor output. Normally several neurons are involved; a single sensory neuron synapses directly with a single motor neuron or a third intercalary neuron links the two together. The presence of a synapse allows the reaction between sensory inflow and motor output to be influenced by information arising outside the reflex arc. The kind of neural activity necessary to produce the common regulatory mechanisms of reflex activity can be illustrated by reference to the control of respiration in insects.

Breathing in insects involves the alternate action of inspiratory muscles which cause air to be drawn into the tracheal system, and expiratory muscles which bring about its expulsion. Correlated with this is the opening of the anterior spiracles and the closing of the posterior ones at inspiration, and the closing of the anterior and opening of the posterior spiracles at expiration. Co-ordination of these movements causes air to be drawn in anteriorly, to flow back through the trachea and to be expelled posteriorly. The system is never inactive but varies both in the frequency and amplitude of the respiratory movements. A locust weighing about 3 grams pumps 120 cc per hour through the system. This can increase to

750 cc per hour although at this rate the spiracles do not act efficiently and much of the air flows tidally in and out of the system.

The rhythm is maintained as follows: within each ganglion of the abdominal nerve cord are a pair of neurons on each side of the mid-line, one of which is capable of firing continuously and innervates the inspiratory muscles of that segment; the other fires only under command from another neuron and innervates the expiratory muscles (Fig. 8.4). It is linked to the inspiratory motor neuron so that when it fires, bringing about an expiratory movement, the inspiratory motor neuron is silent. The immediate command neuron is one which joins the expiratory neurons across the mid-line and so ensures synchronization between the left and right sides of the segment.

The abdominal respiratory movements consist of a unified contraction and expansion. This is accomplished through the action of another pair of command neurons whose cell bodies lie in the metathoracic ganglion and whose axons make synaptic contact with the ganglionic command neuron in each succeeding ganglion. The metathoracic command neuron is capable of firing continuously but is periodically inhibited by cells producing spontaneous oscillatory potentials. Thus the oscillatory firing cell inhibits the continuously firing metathoracic command neuron, introducing in it a rhythmical firing. This causes the ganglionic command neuron to be periodically excited which, by simple transmission, causes bursts of impulses to pass to the excitory muscles; at the same time the spontaneously originated signal to the inspiratory muscles is inhibited (Fig. 8.4).

The control of spiracle activity shows some individual differences in the thoracic and abdominal segments. In general, nervous motor control comes from spontaneous continuous activity of motor neurons to the spiracle muscles. A rhythmical quality is given to these by connections through the nervous system with the metathoracic oscillatory neuron.

Whilst the pumping rhythm is maintained in the complete absence of phasic input, inflow from various sources influences the rhythm to meet the requirements of the animal. Changes in the O_2/CO_2 ratio of air in the trachea and increase in CO_2 leads to hyperventilation of the system. This appears to be brought about by direct action on neurons in the ganglion. Other changes in physiology which require increased ventilatory movements act through the metathoracic neurons in the intact animal. As the ventilatory rate changes with changes in activity, metabolism, temperature, sexual activity, and day/night rhythms of the environment, the pattern of the sensory inflow bringing about change can be very variable and the nervous activity involved in it is complex and variable.

The establishment of this type of reflex relationship where single or a small number of neurons command the integrated responses of complex patterns of activity is widespread in arthropods. In crayfish the escape

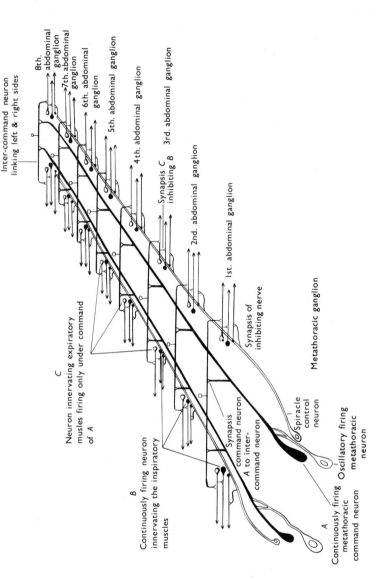

Fig. 8.4 Diagram showing the layout and relationships within the ventral nerve cord of the neurons involved in the control of the respiratory movements of insects.

reaction of darting backwards by flapping the abdomen involves a complex pattern of muscular activity. This is commanded by the giant fibres of the ventral nerve cord which may be activated by pinching the antennae, visual stimulation, or contact with noxious chemicals. More complex responses are found in crabs which can grasp food with the chelae, transfer it to the maxillipeds, taste it, reject it if it is unsuitable or swallow it if suitable, all in the absence of the brain. But, in the latter case, if food is continually supplied to them they cannot stop feeding.

The result of this linking together of neurons is to create a hierarchy so that interactions of higher members of the hierarchy (i.e. system command neurons) can 'ignore' the detail of carrying out their orders, thus simplifying the patterns of reflex activity they would otherwise have to carry and leaving them free to develop higher orders of integration than would otherwise be possible. Neurons concerned with the apex of this pyramid appear to be located in the brain, particularly in the protocerebrum. Often their output appears to be instructions to inhibit the output of systems lower in the hierarchy.

A system designed in this way, with lower members of the hierarchy linked together in definite action patterns by the higher members and containing relatively a rather small number of neurons, tends to give a rather rigid set of patterns. Some modification however is possible, though often only at certain links in the chain, as when the wasp *Ammophila* sp. can determine whether or not to feed caterpillars to her young and how many each should have. Once this has been done on her first visit to the nests no further modification occurs, as for example when the visit is made unnecessary by removal of her larvae.

The association of the input signals with a course of action is fairly flexible and capable of being altered to meet the requirements of the life of the animal. Bees, for example, can learn to link up five differing visual signals with turns in a maze.

In addition to spatial organization certain patterns are linked to occur in a definite sequence. Such patterns as courtship, copulation and oviposition may be programmed so that once the former has started the pattern continues until the last has been completed. Often each phase is signalled to commence and considerable periods may separate them; nevertheless the sequence is maintained and latter events either do not happen or are useless unless the previous ones have occurred. Finally, perhaps without a definite neurological link, patterns may change throughout life to suit the changing circumstances of the growing animal.

INTEGRATION

The next point to consider is how the control mechanisms for the separate events are welded together in the whole animal. Only a simple framework of this very complex relationship can be given here, but without this, the responses of arthropods to changing circumstances will be hard to visualize. The carrying out of any process within or by the whole animal requires the co-ordination of events occurring at all levels of organization and the relating of these events to the environmental conditions under which the act is being performed.

The network of control mechanisms which brings this about shows the following properties:

1. *In situ*, any action is influenced by multiple control mechanisms and, if these mechanisms have an opposing action, fine control to a given value is possible; if they have synergistic actions more rapid change can occur; and finally the controlled event can be changed through diverse paths.

2. Co-ordination is achieved through a hierarchy such that events are influenced by primary control systems, primary control systems by secondary systems, secondary by tertiary ones and so on to higher levels. The number of control systems in any one level decreases as their position in the hierarchy rises, so that a system at a higher level influences several systems at lower levels. Furthermore, at any one level systems may interact together so that a secondary level system could influence a primary system either directly or indirectly, the latter through a more circuitous pathway.

3. Organisms of similar structure and function can vary in the way their control systems are organized by the strengthening of some links and the severing of others. Such variations can lead to diverse responses from quite similar animals. These relationships can be represented diagrammatically (Fig. 8.5) by a truncated pyramid, the broad base of which represents the primary control systems of the animal; above this in the pyramid a plane is drawn which represents the secondary control mechanisms; a third higher up in the pyramid represents the tertiary control mechanisms and the apex the highest control level of the animal, that is, the level at which the most embracing mechanisms of the animal are to be found.

Consider a primary control mechanism operating at (1) and concerned with the regulation of oxygen at the cytochrome level of cellular respiration. Increased metabolic rate in a group of cells leads to increased oxygen requirement. A group of primary mechanisms concerned with membrane permeability and transport within the cell from the adjacent oxygen supply, (tracheolar or pigment), will be activated. This may be sufficient and integration has not proceeded beyond the primary level. Perhaps however, this is not enough to satisfy the increased needs and the supply of oxygen to the vicinity has to be increased by greater blood flow. Then the demand may go to the secondary level and the circulatory system becomes involved. Again, if the needs are met the integration proceeds no further. If they are not met the

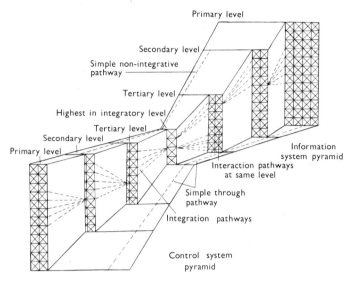

Primary level

Secondary level

Simple non-integrative
pathway

Tertiary level

Highest in integratory level

Tertiary level

Secondary level

Primary level

Information
system pyramid

Interaction pathways
at same level

Simple through
pathway

Integration pathways

Control system
pyramid

Fig. 8.5 Diagram showing the relationships between information inflow and control activity within an animal. For further information see text.

animal needs to increase its rate of breathing and perhaps even to move out of the vicinity of the oxygen shortage. To achieve this more of the body's activities must be co-ordinated and directed to this end and higher levels of the hierarchy activated to bring this about. Indeed the demand may become so great that practically every activity of the body is directed to supply oxygen to the point of need, as may happen when suffocation is induced in the arthropod. Thus the satisfaction of a requirement is ensured within the capabilities of the structure and function of the arthropod by a dynamic system that can extend its influences and ultimately dominate the animal if the need arises, or quietly operate with little reference to the surrounding systems.

In addition to the harmonious working of the animal by these interactions they also have to be related to the environment. This relationship can be expressed by adding to our diagram another truncated pyramid, the apex of which is in contact with the apex of the control system pyramid (Fig. 8.5). This may be called the informational pyramid. Its chief function is to integrate all the information that can be made available to it either from the environment or from the internal state of the animal. The base of the pyramid here represents the primary analysis of stimuli influencing the animal. This analysis is often carried out in the peripheral sense organs of the nervous system. At the next plane is a higher level in the hierarchy of information processing, a single 'message' here being the result of more numerous messages in the primary plane. Tertiary and higher levels may also occur, each handling messages which are the product of more than one element in the next lowest plane. Interactions between events in each plane are also possible. The highest level is identical with the highest level of the control pyramid and here the most integrated information input is related to the most embracing control output.

The information pyramid works in much the same way as the control pyramid: information from a primary source may be fed to other primary sources, augment or cancel each other out and generate no further input, or they may reach to a higher level and, upon relationship with other sources inputting to this level, again be cancelled out or be passed to a higher level. This selection of information is achieved by inhibition of unwanted input at different levels.

This double pyramid model helps to visualize a complex system whose actual anatomy and physiology cannot be so simply expressed. To the model as given must be added the possibility that information at the primary level can be utilized to affect a control mechanism at the primary level without passage to the higher levels of the hierarchy. This can and does happen but the diagrams to express all these possible pathways would be as complex as the details of the anatomy of the animal which contain the pathways actually taken.

9

The Evolution of Arthropods

INTRODUCTION

Any member of the present-day arthropod fauna is a product of the interactions between many generations of individual animals, each of which represents a variation of arthropod design, and the environments in which they have existed. This interaction has been going on for at least 500 million years; it is happening now and will continue into the distant future. In the past the original arthropod population has, by a process of phylogenesis (speciation), given rise to a vast number of forms. Some of them were so successful that their design dominated the further development of the group. Others endured only for a limited period of time before they became extinct. This aspect of arthropod existence will be dealt with by considering the nature of the programmes which were tried out at each generation against the environment and the processes of evolution by which the numbers and kinds of programmes occurring in a population are determined. This is followed by the manner in which species arise from existing species and leads on to a history of the dominant trends that can be distinguished within the arthropod stock and their relationships to each other. Finally, there is the problem of expressing these relationships by a system of classification which will be briefly described.

THE NATURE OF THE GENETIC PROGRAMME

The genetic programme of the animal comprises a set of paired genes, one member of the pair inherited from one parent and the other from the other parent. The entire complement of genes, which may amount to five thousand pairs, is carried on the chromosomes within the cell nucleus. Initially there is the single zygote nucleus but in development this is replicated many millions of times. The entire gene content of the animal at any one time is the genome. The replication of the genetic programme is not necessarily exact. Deviation from strict replication may be part of the design of the animal.

Once replication has occurred the gene content of a cell can differ from its fellows by endomitosis in which there is replication of the chromosomes without nuclear division. In insects this gives the clear banded patterns seen in Diptera where, in the salivary glands, replication may be 1064 times (Fig. 2.1). The value of this is to give an increased number of sources of mRNA formation for any one DNA instruction, thus increasing the speed and amount that can be formed from it. The number of endomitotic replications is characteristic of the type of tissue of which the cell is a member.

While the genetic programme content of the cells appears never or only in certain exceptional cases to fall below that of the original zygote, the same information is not called forth from it in each cell. In Diptera it can be seen clearly where the puffs on the chromosomes which signal gene activity form different but characteristic patterns for each cell function (Figs. 2.2 and 2.3).

The alleles present may not be identical. One member of the pair may have information to produce an isoenzyme of the product of the other. This isoenzyme may be called into play in some tissues of the animal, while another form is used in other cells (malic dehydrogenase in the moth *Cynthia*). In esterases in *D. melanogaster* one form is quickly inactivated by heat (50° C) while another shows little loss of activity at this temperature. Alleles may contribute information necessary for normal functioning or give the basis for extending its performance when stress exceeds the capabilities of one allele product to cope. The product of a gene appears to be a protein of a specific type. The appearance of this protein in the right place, at the right time and in the right amount is essential for the functioning of the animal. Its failure to appear whenever required (which may well be more than once and in different places) will cause malfunctioning of the animal. Early in development this can be serious and lethal to the animal. Later it may only lead to small malformations and/or sterility.

The programme of the animal has been defined as the information contained within the zygote which in decoding and interacting with the environment produces the individual arthropod. This programme contains a large amount of information and it is convenient to divide it into four sub-programmes whose contents can vary independently of each other.

The cellular and biochemical (C.B.C.) sub-programme

This contains information which sets up the general cellular structure and biochemical reactions of the animal, which arthropods share with other

Phyla. The microstructure of the cell, the biochemical pathways of protein, carbohydrate and fat metabolism, and the mechanisms of ionic balance and of membrane permeability belong here. The events controlled from this programme will occur throughout the life of the animal at cellular levels. The mRNA derived from the DNA base of the programme is produced as economically as possible; as much of the replication as can be is handed over to the cytoplasm.

The general arthropod design (G.A.D.) sub-programme

This contains the information necessary to organize the growing and developing animal into an arthropod of some particular kind. This may include information about the synthesis of specific proteins where they are peculiar to arthropods and important to their design. It is however primarily a sub-programme which organizes events in space and time, playing upon the basic properties of the **C.B.C.** sub-programme so that the multiplying cells take up the special configurations of arthropod organization, produce their products and assume their functions in accordance with this and no other design. Where the life pattern is a succession of different forms this sub-programme contains information for the production of these forms and in their proper sequence. The reading of this sub-programme is highly organized and the production of mRNA from the DNA codings is closely controlled, occurs in definite sequence, and often each coding is active for only a strictly limited period.

The function, range and homeostasis (F.R.H.) sub-programme

This sub-programme contains the information which dictates the kinds of materials which organs process and/or the stimuli to which they respond, e.g. the presence or absence of colour vision in compound eyes; the ranges of physical conditions (temperature, humidity, ionic concentrations, etc.) over which the animal can function normally; and the ability of the regulatory mechanisms to maintain the body functions in harmony with each other under diverse environmental conditions.

This sub-programme, although not as homogeneous as the others, forms a unit because it is in these features that closely related populations and species most often differ from each other. In addition, repetitions of the variations that occur in this sub-programme tend to be repeated again and again in groups differing widely in **G.A.D.** sub-programme content.

The alternative and redundant information (A.R.I.) sub-programme

In some respects this is the emergency sub-programme of the arthropod and is composed of two sets of information: (1) the alternative information, a set of instructions that may be called into operation instead of another set should the circumstances demand it.   An arthropod which contains instructions for the diapause state, but may or may not enter it, depending upon climatic conditions it encounters, is an example.   (2) The redundant information results in a single process having more than one mechanism to ensure its functioning, each being capable of doing the whole job should the need arise.   Under these circumstances failure of one mechanism would not be fatal to the animal, and the effects of error, always a possibility in such a complex situation as a developing animal, are guarded against.

THE EVOLUTIONARY HISTORY OF THE ARTHROPODA

Onychophora—Myriapoda—Insecta (Fig. 9.1)

Amongst living animals the Onychophora most nearly resemble in general body form, uniformity of segments, simple head structure, possession of lobopod limb, and manner of locomotion, our concept of what the earliest Arthropoda were like. They live on the surface of the soil or amongst vegetable debris on the forest floor. Movement through this type of habitat, particularly the ability to squeeze through narrow crevices, is good reason for the persistence of these structures and the animals show no evidence of secondary simplification in their general structure. Onychophora can be traced back to Mid-Cambrian Protonychophora which lacked slime papillae and jaws and lived in a marine habitat. In present day Onychophora adaptation to terrestrial conditions is expressed in the dry urate excretion from the gut, internal fertilization, and in most species the production of viviparous young. These adaptive features are at a primitive level: the animal cannot resist dessication and is restricted to damp localities. The spermatophore is attached to the integument, the sperms entering the haemocoele and thus passing to the spermatheca.

A comparison of the living arthropods with the Onychophora singles out the Myriapoda as most closely resembling them. The major differences lie in the development of a more complex head which is distinct from the trunk segments, the latter developing as a unit and not showing further tagmosis. The cuticle has developed into an exoskeleton and the limbs have elongated and taken on a complex articulated structure.

Persistent resemblances are found in the simple mid-gut without diventricula, the elimination of uric acid crystals from the gut and the presence of exertile vesicles. Some characteristic embryonic features such as the arrangement of the coelomic vesicles, the point of origin of the ventral ganglion and the presence of 'ventral organs' are features common to both the Onychophora and Myriapoda.

The earliest records of the Myriapoda are from the Upper Silurian period and are roughly contemporaneous with the earliest land fauna.

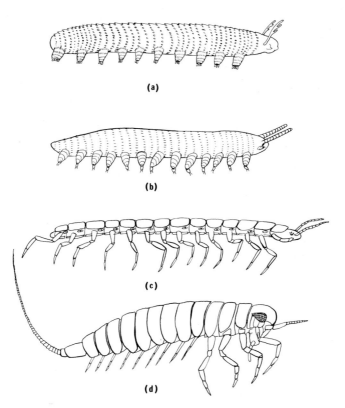

Fig. 9.1 Onychophora—Myriapoda—Insecta evolutionary trend. (a) Protony-chophora, *Aysheaia*, Cambrian. (b) Onychophora, *Peripatus*, modern. (c) Myria-poda, Chilopoda, modern. (d) Insecta, Monura, Carboniferous Period. (Redrawn from Sharov 1959)

The head structure in the Myriapoda consists of a fusion between the anterior prostomium and the first three body segments, the pre-antennal, antennal and pre-mandibular units, which have come to lie in front of the mouth. Externally in the adult animal the pre-antennal segment is in-distinguishable; the antennal segment is marked by the presence of a pair of antennae, and the pre-mandibular segment sometimes by the opening of its coelomoducts.

Within the Myriapoda other segments (gnathal segments), arc present in the head but are not pre-oral in position. The mandibular segment and a maxillary segment, the latter with a 'lower lip' between the maxillae, complete the head in the Pauropoda (Dignatha), while in others

(Trignatha), three gnathal segments are present, mandibular, maxillary and labial. In the Chilopoda the appendages of the first trunk segments have evolved to form poison claws which project forward under the head and virtually conceal the mouthparts.

In their trunk segments the Myriapoda lack tagmosis but the trunk should be thought of as a unit and not as a series of segments.

In their general body form the Insecta are a very distinct group of Arthropoda. The head, formed by the prostomium and six segments, is a closely welded capsule bearing compound eyes and ocelli, a single pair of antennae, and three gnathal segments, the mandibular, maxillary, and labial. Relative to the trunk the head is mobile and there is a distinct neck. The trunk shows tagmosis, and the three segments immediately following the head are specialized for locomotion, each bearing a well-developed pair of walking legs. The remainder of the trunk of eleven segments lacks locomotory appendages although a pair of slender articulate cerci and a single medial anal filament may be present in primitive forms.

A marked characteristic of many insects is their ability to fly, wings being developed upon the second and third segments of the thorax. Wings however, were developed within the group and a number of primitive wingless insects may be found widely distributed amongst living faunas throughout the world.

Compared with Myriapoda the main advance in insects has been the elongation of the limbs conferring longer strides and greater speeds, a reduction in the number of limbs eliminating mechanical interference and allowing the execution of a greater number of gaits, and the countering of the increased tendency to undulations of the body which these changes would bring about, by having the bases of the three pairs of limbs set close together on a rigid plate giving the characteristic hexapod condition.

These locomotory advantages may have evolved more than once since we find considerable structural diversity amongst the Orders exhibiting them. Of these Orders, the Collembola (Springtails), are an important element in present-day microfaunas. Fossil Collembola are known from the Middle Devonian period from Rhynie Chert in Northern Scotland. Apart from differences in the antennae the animal shows all the characteristics of present-day forms. The Protura are known only from recent forms: they are minute, lack antennae and show an increase of segment number from eight to twelve during growth. They appear to be very common in vegetable debris and are predaceous. The Dipleura are known from the Baltic Amber of the Oligocene Period and from recent forms.

The Thysanura are present as fossils in the Lower Oligocene Baltic Amber and there is a widespread modern fauna. This group closely resembles the Pterygota insects and the Myriapoda in the uniform build of the head, the mandible formed from the whole limb, the presence of the organ of Tömösvary, uniramous walking legs with exertile vesicles at their base, two surviving segmental organs in the head, the pre-mandibular and first maxillary, the simple gut, the elimination of solid excretory particles, the presence of Malpighian tubules and the close resemblances in the structures of the tracheal system, the heart, the gonads and their exit ducts.

From the Lower Permian of S.W. Siberia some specimens of the genus *Dasyleptus* belonging to an Order Monura have been described. These closely resemble the Thysanura in general form but show many more primitive features; in the head the tergites of the mandibular and labial segments can be distinguished as separate plates; the fourteen thoracic and abdominal segments form an even series; the thoracic segment bearing simple limbs with undivided tarsus and claw is not enlarged; nine pairs of short abdominal limbs are present and the last segment bears a long median style.

These Orders of the Collembola, Protura, Dipleura, Thysanura and Monura are sometimes classified as insects in the Sub-class, Apterygota, and sometimes only the Monura and Thysanura are so grouped, the remainder being placed in the Myriapoda. In any case the Monura and Thysanura are closer to the pterygote insects though probably not to their immediate ancestors.

The winged insects are classified as one Sub-class, the Pterygota, and while many members of the group lack wings it is clear that this is a secondary condition. As revealed by fossils preserved in coal measures, an extensive fauna of Pterygota was present in the Middle Carboniferous Period. It was already diverse and must have been preceded by insect faunas in which wings were evolved. The fossil record offers no intermediate forms which help to solve the stages and circumstances which lead to the evolution of flight in insects.

Wings are only functional as organs of flight in adult insects and preserved nymphal forms from the Carboniferous indicate that this is a primitive feature. In these primitive forms, the wings developed in ontogeny very much as they do in present-day insects. The growth and development of a wing once it had become functional probably presented many difficulties. One of the stone flies, *Atactophlebia termitoides*, from the Lower Permian appears to have moulted twice in the adult stage, judged from the structure and venation of its wings. In modern Ephemeroptera the cuticle is delaminated and shed shortly after the adult form is reached. This, however, appears to be an adaptive device to lighten the body in flight. It has evolved within the group and delamination of the outer cuticular layers is a very different procedure from the mitosis and new cuticle formation associated with true growth and moulting.

The earliest pterygote fauna of which we have any knowledge lived in the hot and humid climate of the tropical forests that stretched across Canada in Middle Carboniferous times. Their remains now found in the coal measures indicate a sudden appearance in great numbers, diversified in form and in different areas of Laurentia, N.E. America and W. Europe at the same time. The origin of these is unknown but presumably had occurred elsewhere during the preceding period.

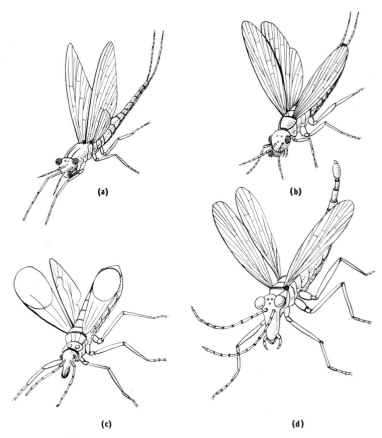

Fig. 9.2 Insecta, Pterygota. (**a**) Ephemeroptera. (**b**) *Perlaria*. (**c**) Hemiptera. (**d**) Mecoptera.

Differences in wing structure show that two major types had evolved. In one, the Palaeodictyoptera, the wings were held out horizontally and could not be folded back over the abdomen when at rest but could be folded vertically above the thorax; both pairs were similar and the hind pair lacked a jugal field. In the other, the Polyneoptera, the forewings were specialized as tegmina and the hind wings broadened posteriorly to form a jugal lobe which greatly added to the lifting power of the wing in flight. Immature forms of both groups are known. In the Paleodictyoptera fossil imprints show the presence of compound eyes and rudiments of wings sticking out at right angles to the body. There is some evidence that they were aquatic and underwent a great many moults in the water before coming out as a sub-imago to undergo a final moult on dry land. In the Polyneoptera the forms were very like young cockroaches of the present day but the wing pads were longer as were also the cerci. In some these were flat and broad and are thought to indicate an aquatic habitat.

The adults of these early Pterygota were very big, differing from the young forms only in the possession of wings. They differed from modern insects in the absence of all the complications of shape and form, reduction and loss of structures that have come about through specialization of their anatomy. The head was typically insectan; on the thorax the three pairs of legs were identical; the wings were alike and, of the eleven abdominal segments, the first ten were identical. Some were phytophagous and some carnivorous. The whole of this fauna, which reached its peak in the Stephanian Period, gradually became extinct throughout the Permian Period as the hot marshy conditions were replaced by cool, more arid, conditions.

Prior to this happening some stocks spread out over the land in the Devonian or Carboniferous Period and eventually entered Gondwanaland. Here, under temperate climatic conditions, further evolution of the Palaeodictyoptera and Polyneoptera occurred and two further major groups of insects, the Paraneoptera and the Oligoneoptera, arose.

The present-day mayflies (Ephemeroptera) share several features with the Palaeodictyoptera. The wings cannot be folded horizontally over the abdomen, the muscles are attached directly to the bases of the wings and the abdominal segments remain homonomous. It is thought likely that the group arose in Gondwanaland at the end of the Carboniferous Period, the early (Permian) forms having two pairs of similar wings and well developed mouthparts. In modern (Mesozoic) forms the mouthparts are degenerate in the adult and the hind wings greatly reduced or absent.

Modern-day dragonflies, Odonata, also show features that link them to the Ephemeroptera and Palaeodictyoptera. One archaic group of dragonflies arose in the hot Laurentian forests in Carboniferous times adapting itself to increasing aridity and dying out in Jurassic times. Present-day dragonflies arose in Gondwanaland during the Permian Period, modern sub-orders being recognizable in the Jurassic of Upper Bavaria.

Of the Polyneoptera the cockroaches appear to have remained unchanged to the present day, little difference being visible between Carboniferous and modern forms. In detail however, all the Paleozoic families had disappeared. The stone-flies appear in the Permian in Gondwanaland and almost all the evolutionary lines have direct descendants still alive. They were aquatic and have histories similar to those of the mayflies and dragonflies. The longhorned grasshoppers were represented in the Upper Carboniferous.

Of the two new groups that arose in Gondwanaland, that of the Oligoneoptera have a highly evolved jugal area on the hind wing with one single vein: the wings are often linked together by a structure ensuring that they move together and present a single surface to the air. In most cases the Malpighian tubules are few in number. In their life history a quiescent pupal stage occurs between the last immature form and the adult. Such a stage is thought to have arisen as a resistant form to the cooler winters of temperate Gondwanaland. In such a quiescent stage freed from the need to combat its environment, considerable morphological change is possible which permits the immature forms to exploit habitats and food sources different from those of the adult, thus reducing competition and permitting a much larger population to survive. This feature must have contributed greatly to the group which, containing the alder flies, scorpion flies, beetles, lace-wings, butterflies, moths, fleas, ants, bees, wasps, ichneumon flies, sawflies, and true flies, probably comprises the bulk of today's insect fauna.

The beetles (Coleoptera) arose in the Permian Period and the majority of the wood-feeding, carnivorous and saprophagous groups that are present among the beetles today had already appeared by the Jurassic Period. The phytophagous, leaf-living and flower-feeding groups did not appear until the Cretaceous times and the lace wings and alder flies (Neuroptera) arose in Permian times.

Merostomata—Arachnida—Trilobita (Fig. 9.3)

The fossil fauna of the Lower Cambrian shows many highly organized arthropods which lack any indication of near-annelid origin. Members of the Crustacea, Trilobita and Chelicerata are present, together with survivors of what must have been an early Paleozoic fauna. This latter contains forms which are unquestionably fully organized arthropods but do seem to lie near the base of the subsequent crustacean, chelicerate and trilobite developments.

The Trilobita, (Fig. 9.3), were the dominant arthropods of the early Paleozoic seas and survived until Permian times. The body consisted of a cephalon with segmentally furrowed glabella, paired sessile compound eyes and a large ventral labrum. Behind the cephalon was a segmented trunk ending in a pygidium. Large pleural plates projected laterally covering the

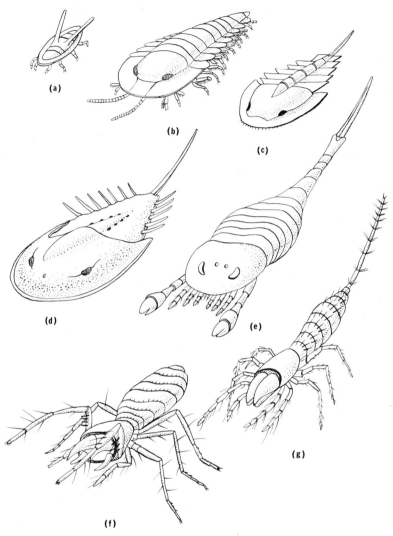

Fig. 9.3 Trilobita, Merostomata, Arachnida. (**a**) Protaspis larva of a trilobite. (**b**) Adult trilobite. (**c**) Trilobite larva of *Limulus* (Merostomata.) (**d**) Adult *Limulus*. (**e**) *Eurypterus* (Merostomata) Ordovician—Permian Period. (**f**) *Solifugae* (Arachnida) recent. (**g**) *Palpigradi* (Arachnida) recent.

limbs. Appendages are known from only about 10 genera. Anteriorly there was a pair of antennae and on each of the successive segments a pair of biramous limbs. Those of the cephalon were crowded together, the coxae projecting towards the mid-line but apparently not meeting close enough to make grinding and crushing between them possible. Those of the trunk segments lie far apart. The animals are judged to be bottom dwellers and possibly were mud and suspension feeders. The outer ramus of the biramous limb carried a broad fringe of filaments. The gut showed in one case two pairs of caeca. The larval form of *Protaspia* was planktonic and grew in size by multiplication of segments in front of the most posterior one.

Compared with other arthropods the group was remarkably conservative in structure and its failure may have been due to its inability to develop fast enough movements to escape its more active predators such as fish and cephalopods.

A number of living forms of arthropods such as the scorpions, king crabs, spiders, pseudoscorpions, harvestmen, ticks and mites, have the first pair of appendages modified to form chelicerae which are used for grasping and thus contrast greatly with the sensory antennae of other arthropod groups. The work of Lankester on the king crab, *Limulus polyphemus*, indicated that these were a monophyletic group and that the possession of chelicerae characterizes one of the great natural assemblages of arthropods, the Chelicerata.

Within the Chelicerata the trilobite larvae of the king crab (Fig. 9.3), is the only arthropod outside the Trilobita to show this type of body structure. The limbs show features also found on the trilobite limb: both groups have branching mid-gut diverticula and lack true jaws. Some fossil forms, *Emeraldella* and *Habelia* present in the primordial arthropod fauna have features suggestive of both trilobites and merostomes (Merostomata, comprising the fossil Eurypterida, and the modern king crabs). Trilobites however have antennae as their first appendages. These are thought to have developed a chelate structure (many antennae in other groups have adopted a grasping function). Further modification would lead to chelicerae with the appropriate modifications of feeding and habitat this would bring about. Such changes must have occurred very early in the history of the Trilobita. Indeed the Trilobita, as represented in the fossil fauna, could not have been ancestors of the Chelicerata, but the derivation of the latter from a pro-trilobite stem is acceptable.

The earliest fossils which show undoubted chelicerate features are the Aglaspida from the Lower Cambrian to the Upper Ordovician, and the Eurypterida from Ordovician times. The former were bottom dwelling marine animals with typical prosoma-opisthosoma tagmosis. The prosoma was covered by a shield bearing a pair of prominent unfaceted eyes and bore ventrally six pairs of limbs, the first pair being chelicerae. The opisthosoma was of eight simple segments each bearing a pair of appendages. The five pairs of appendages of the prosoma and the eight pairs of the opisthosoma were simple uniramous walking limbs. The Eurypterida (Fig. 9.3) were free-swimming animals; the early forms were marine but later forms are found mainly amongst brackish or freshwater faunas. The prosoma had medium dorsal ocelli, a pair of lateral eyes and ventrally six pairs of appendages, the anterior ones being chelicerae, the others specialized for grasping, swimming and walking. The opisthosoma had twelve segments and ended in a terminal spine. The anterior five segments of the opisthosoma had five pairs of gills associated with their sternal plates. The small larval forms had very large eyes, few body segments and bore no resemblance to the trilobite larvae of the king crab. The modern king crabs (Xiphosura) are marine survivors of this once extensive merostome fauna. They are best known from the species *Limulus polyphemus* which is a bottom dwelling inhabitant of shallow coastal seas and estuaries. The animal is adapted for creeping through mud and sand and for feeding upon the soft worms and hard lamellibranch molluscs that live there.

The prosoma is large and its dorsal plate (carapace, but not to be confused with the carapace of Crustacea) is expanded laterally and anteriorly to form a hood which covers the limbs: the mouth lies far back from the anterior edge on the ventral side. Dorsally the prosoma bears a pair of eyes. In the opisthosoma the tergites are also fused to form a single dorsal plate which is hinged to the posterior edge of the prosoma. The opisthosoma is short and has a long strong spine hinged to its posterior border.

Ventrally in front of the mouth is the median labrum and the pair of chelicerae. The six pairs of limbs of the prosoma bear chelicae at their tips and have strong gnathal processes on their coxae. The limbs are grouped around the mouth, the gnathobases directed towards it, giving an almost radial symmetry to the limb distribution. The opisthosoma has six pairs of appendages. The first form a pair of plates, the genital operculum (chilaria), covering the genital openings. The remaining five pairs form plate-like paddles used when swimming and they also bear gills on their inner surfaces. Together the plates form a respiratory chamber, the outer surface protecting the delicate gills from clogging and being damaged by the mud and sand in which the animal lives.

The limbs of the prosoma are used for walking, gathering and chewing or crushing food. In walking, the animal utilizes the anterior posterior swing of the limb: if the limb tips come into contact with food they grasp it in the chelae and convey it to the gnathobases. Soft food is manipulated by the action of the gnathobases of limbs 3–5 which now work by a transverse movement of the limb base to shred the food into fine pieces, the fragments being conveyed to the mouth by the chelicerae. Since walking and feeding require different and incompatible movements of the prosomal limbs, the animal cannot do both at the same time. (Compare with the multiple functions of the crustacean limb.) When hard food is collected (lamellibranchs) the mollusc is gripped between the chilaria and crushed by the biting action of the strong large gnathobases of the sixth pair of prosomal limbs. The soft material is then shredded further by the bases of the more anterior limbs and passed to the mouth by the chelicerae. Quite coarse particles enter the mouth but these are ground into smaller size by the gizzard so that only fine particles enter the mid-gut.

The remaining Chelicerata are grouped together as the Arachnida. Unlike *Limulus* and perhaps the other Merostomata, the Arachnida are all fluid feeders although they chew their prey with the coxae of the pedipalps for long periods. In the scorpion the prey is caught by the chelae of the pedipalps, crunched several times, brought to the chelicerae which tear off pieces and hold them to the pre-oral cavity. Digestive fluid from the mouth is alternately poured over the food and sucked back.

Within the groups that make up the Arachnida, evolutionary relationships are made very complex by the presence of primitive characters in the higher groups, some of which appear more primitive than those of the Merostomata.

The fossil record shows that the earliest Arachnida appeared in the Silurian period and were the scorpions. Almost all the higher Orders appear abruptly in the faunas of Devonian and Carboniferous times and under terrestrial conditions adverse to their fossil preservation. Amongst terrestrial arthropods only the Silurian scorpions show any resemblance to known aquatic chelicerates. The Silurian *Palaeophonus* is found in association with eurypterids and is thought to be an aquatic animal, but it could have led an air-breathing existence along the debris line on the sea shore. Of the scorpions from the Carboniferous Period some show true stigmata while others do not. In modern scorpions the lung books are formed not by invagination of the stigmata but by the bending back of embryonic limbs, each covering over a recess along the hind margin of the limb base. The first lung lamellae arise from the hind surface of the limb. Lankester advanced the theory that the lung books of scorpions were homologous with the gills of the merostomes. The manner in which they are formed during early development and in which they are enclosed in a chamber by the fusion of the plate with the body wall confirms this point.

The chelicerates were one of the earliest of arthropod groups to invade the land. Tracheae to replace the lung books as respiratory organs evolved in several lines within the group. Other adaptations to terrestrial existence are shown in their efficient water conservation mechanisms, dry nitrogenous excretion (mainly guanine), and widespread occurrence of viviparous reproduction. As a group they established a stable tagmosis pattern of prosoma and opisthosoma which then specialized in many lines and ways by simplification of the original plan, as shown by the reduction of the opisthosoma encountered so frequently in the different arachnid groups.

The Crustacea

Living representatives of this group (Fig. 9.4), the Branchiopoda, Ostracoda, Copepoda, Cirripedia, and Malacostraca, are widespread in aquatic habitats throughout the globe, both in marine and fresh water. Terrestrial forms, although commonly found, consist of only a small and

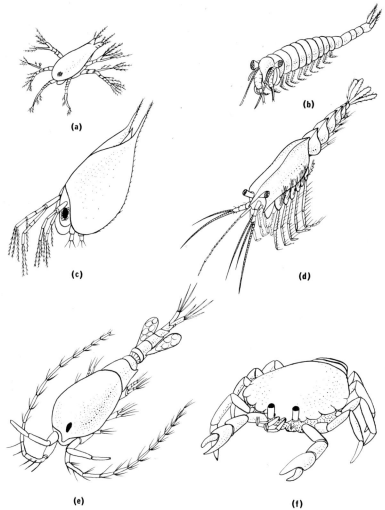

Fig. 9.4   Crustacea. (a) Nauplius larva of *Penaeus*. (b) *Chirocephalus* (Branchiopoda). (c) *Daphnia* (Branchiopoda). (d) *Euphausia* (Malacostraca). (e) *Calanus* (Copepoda). (f) Brachyura (Malacostraca). (For Cirripedia see Fig. 10.3)

rather specialized number of species. The groups named illustrate the diversity of the Crustacea, while the possession of nauplius larvae (Fig. 9.4), in all groups and the presence of pre-oral antennules and antennae indicate that these are of monophyletic origin.

The limbs of Crustacea indicate that the primordial animal had a series of similar biramous limbs. These differ in structure from those of the trilobites but the mere presence of biramous limbs in both groups may indicate some affinity between them.

The nauplius larvae of Crustacea and the adults of Cladocera, Conchostraca swim, as did the Devonian *Lepidocaris*, with their antennae, probably the primitive method. Further swimming organs developed on the thorax and abdomen in the adults of other groups. Rapid escape reactions, resulting from the simultaneous movement of thoracic limbs (2nd and posterior) in copepods and of the whole abdomen in decapods, are a further specialized feature requiring the development of a giant fibre system for its accomplishment.

The general body form of a crustacean is not that of a bottom dweller but is slender and flexible with movable pedunculate compound eyes. Swimming, using limbs, appears to have been primarily a matter of paddling, which permits swift efficient movements in animals of the size of a water flea (*Daphnia*), but gives only slow progression to larger animals such as a fairy shrimp (*Chirocephalus*). These Crustacea survive only where large predators are absent, such as in temporary waters, or possess a fecundity and rapidity of growth which enables them as a species to stand a heavy predation rate. Larger animals may also survive in competition with fast moving predators by the development of swift and competent escape reactions or by having a heavy exoskeleton as found in large slow growing living animals. Very small animals do not find paddling an efficient means of locomotion and the very tiny Crustacea which occur in water films move by crawling rather than by swimming.

Suspension and detritus feeding is common throughout the Crustacea. Filter feeding mechanisms differ fundamentally in detail from group to group and are the result of independent evolution. In Branchiopoda and Malacostraca the interlimb spaces on each side of the body expand and contract as the limb moves, causing water to be drawn away from the mid-ventral line where food particles are strained from the outward backwardly directed stream by setae set on the basal endites of the limb. Food thus strained may be carried forward by orally directed currents or by the mechanical action of setae transferring the food to the more anterior limb. The food is helped into the mouth by the mandibles.

In the copepods the maxillae bear a single pair of filter plates which do not move; the water current is created in swimming forming a vortex about the head and limbs. The Cirrepedia employ feathery legs as casting nets which scrape themselves clean as they curl up, the food passing to the mouth.

Large food feeding also occurs commonly throughout the Crustacea even in animals which are generally filter feeders. This ability has helped the animals to increase their size above the limit of filter feeding maintenance.

In *Chirocephalus* the mandibles are essentially rubbing, grinding and squeezing organs and do not exert a cutting action. This action depends mainly on the rolling of the mandible antero-posteriorly about a dorso-ventral axis, action which resembles the promotor-remotor swing of the coxa of a walking leg and has presumably been derived therefrom. The development of a transverse bite has occurred many times in the Crustacea and details differ in the different groups.

The Pycnogonida (Fig. 9.5)

These are the sea spiders which are widespread in all the marine habitats. In the adult form the anterior limbs form a pair of grasping organs, the chelophores; the next pair, palps; and third pair, ovigers to which the eggs may be attached. The trunk bears four pairs of long walking limbs: a tiny non-segmented region terminates the body. Paired multiple gonopores are present on the trunk segment, the only case in arthropods where more than a single pair of gonads and their ducts occur. Associated with the ventral nerve ganglia are paired ventral organs, found elsewhere in the arthropods in the Onychophora. The body form is thought to be functionally related to the needs of crawling amongst the hydroids on which the animals feed, either grasping food with their chelophores or, where these are absent, feeding by a specialized proboscis which is used as a suctorial organ. The earliest Pycnogonida are known from the Devonian faunas.

The Tardigrada (Fig. 9.5b)

One other group of arthropods is known which shows features suggesting derivation from some basal stock. The Tardigrada are minute animals with an anterior terminal mouth, completely lacking a pre-oral cavity. Their appendages are simple lobopod type and end in claws. The nervous system is of primitive ladder type. They are tiny animals found in moss and damp places throughout the world.

THE ORIGIN OF ARTHROPODS

The evidence presented in previous chapters indicates that those animals included in the Phylum Arthropoda show considerable differences in their organization and that these differences are already present in the simpler more primitive faunas of the Cambrian Period. The question arises whether

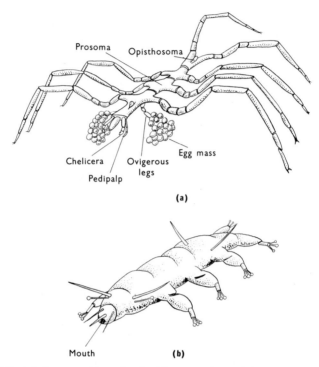

Prosoma

Opisthosoma

Chelicera

Pedipalp

Ovigerous legs

Egg mass

(a)

Mouth   **(b)**

Fig. 9.5 (a) Pycnogonida male. In this group the males carry the eggs. (b) Tardigrada, a minute arthropod whose limbs resemble those of *Peripatus*. Both these groups are difficult to relate to the evolutionary patterns of the the rest of the arthropods.

the arthropods are monophyletic, that is, have arisen from an animal that was itself an arthropod; or polyphyletic, that is, the features which they share in common and which distinguish them from all other Phyla have been acquired independently in several evolutionary lines arising from segmented worms.

While a final answer cannot be given at the present time the following evidence bears upon this problem. The segmental plan of body structure consisting of coelom, coelomoducts and nerve cord has been developed in the Annelida, the Chordata, and perhaps in the molluscs and hence it is not unique. Cephalization, that is specialization of the anterior pre-segmental region and segments to form a head tagma, occurs independently in each group. While the Annelida represent the type of segmented worm from which arthropods probably arose, it is not so certain that they are survivors of actual lines that gave rise to the arthropods.

The new features which appear in arthropods and which make them superior to segmental worms are the firm exoskeleton coupled to growth by ecdysis, jointed appendages, striated muscle, haemocoele and ostiate heart. These specializations are of considerable functional importance and could have arisen more than once. If this is so, then the term Arthropoda is really a name for a grade of organization and not of a unique type of organization. It is worth remembering that the basic organization of an arthropod is that of a segmental worm and not a pattern of a different type.

THE CLASSIFICATION OF ARTHROPODS

The value of comparative reasoning in biology depends upon the correct expression of the relationship between the organisms being compared. These relationships are best expressed through a system of classification which must be phylogenetic, that is express natural relationships. Other systems which are often aimed at aiding identification have their uses but do not provide a basis for evaluating the resemblances and differences that comparison reveals. The relationships between the groups of arthropods have been indicated in the evolutionary account just given, which, while it represents a simple version and is as accurate as this allows, is certainly not the final verdict. This will not be given until a great deal more knowledge on all aspects of arthropod biology is available. Meanwhile other versions are given by other workers and to the confusion of the student scarcely any two textbooks agree on the major classification of this group. The following paragraphs aim to set forth a reasonable classification of the group and to put into perspective the difficulties and reasons for the diversity that exists.

There is perhaps greatest measure of agreement on the contents of each of the arthropod groups which rank as Orders in the systematic hierarchy. It would be reckless to try to define an Order but survey of the many descriptive texts and natural history books indicates that Orders are a fairly distinctive grade of body organization: for example, the Isopoda (woodlice), Scorpiones (scorpions), Lithobiomorpha (centipedes), Lepidoptera (butterflies and moths), and Xiphosura (king crabs). A more extended list of arthropod Orders is given on page 190.

Similarities in pattern show that some of the Orders are more closely related than others: the Orders of similar design are grouped together to form Classes. The Classes of arthropods are: Onychophora, Myriapoda, Insecta, Trilobita, Merostomata, Arachnida, Crustacea, Pycnogonida and Tardigrada. The number of Orders within a Class may be as low as one in the Onychophora, two in the Tardigrada, or, amongst living forms, as many as 29 in the Insecta and 35 in the Crustacea. Where a number of Orders are present in a Class, degrees of greater or lesser resemblances can be recognized and need to be expressed. Seven Sub-classes are recognized in Crustacea and two in Insecta, while in Arachnida the eleven Orders present are sufficiently different from (or similar to) each other that no ranking above the Order level seems acceptable.

Amongst the Crustacea the largest Sub-class, the Malacostraca, contains six groups of Orders called Super-orders: the first of these, the Super-order Leptostraca, contains only one Order, the Nebaliacea, the most primitive member of the group, while the larger Super-order Peracarida comprises seven Orders expressive of the wide radiation here present.

In the Insecta, four or five Orders have never developed wings and are grouped together as the Sub-class Apterygota, while the other Orders have wings or had them and are grouped together as the Sub-class Pterygota. Within the Pterygota, relationships between Orders have been expressed in different ways. Modern studies favour the idea of four groupings called Divisions indicating four separate and equal radiations into diverse designs. The first of these forms the Division Palaeoptera: it contains the ancient Order Palaeodictyoptera which appears to have been the parent group for all winged insects, together with the modern mayflies (Ephemeroptera) and dragonflies (Odonata). The other Divisions are the Polyneoptera (cockroaches, white ants, grasshoppers, etc.); the Paraneoptera (the bugs and lice); the Oligoneoptera (the beetles, butterflies, true flies, fleas, wasps, etc.). The latter group of Orders differs in several other prominent features from the rest, principally in the wings which, in the immature stages, develop locked inside the larval epidermis. Hence the Division Oligoneoptera is sometimes referred to as the Endopterygota, all the others then being treated collectively as the Exopterygota, as the wing rudiments grow exposed from the thoracic surface. Some workers have been impressed by the different life cycles found in insects which roughly correspond with the grouping Apterygota, Exopterygota and Endopterygota. They refer to them as the Ametabola where metamorphosis is absent, Hemimetabola with partial metamorphosis, and Holometabola with total metamorphosis between the last larval stage and the adult form.

The assembling of the Classes of arthropods into higher categories to express the different degrees of similarity between them has produced many different systems of classification. Present-day studies emphasize the close relationships of the Onychophora, Myriapoda and Insecta as one grouping; the Merostomata, Arachnida and Trilobita as another; and the Crustacea as a third distinct group. Both the Pycnogonida and the Tardigrada stand apart from these groups and from each other. They may have relationships with a pre-onychophoran arthropod stem. No names seem to have been given to these three groups and in view of the multitude of names given to other groupings it is perhaps as well to leave these relationships unnamed until their status is unquestionably established.

One or two other systems need to be briefly mentioned:

1. Those such as Haeckel's early work which associates the Orders into two groups based on the possession of a character which it was reasonable to suppose evolved only once, such as the gills and the trachea. All terrestrial arthropods with tracheae formed the Tracheata, aquatic ones with gills, the Carides. Later, with new knowledge about *Peripatus*, Haeckel proposed the Tracheata and the Crustacea, expressing doubt that they could have a common origin and suggesting that the arthropods were a diphyletic group.

2. Those such as Lankester's work which was founded on the number of segments that assumed a pre-oral position and came to be incorporated into the head. A group, the Hyparthropoda, was proposed for cases where no pre-oral segment had developed. No such creature is known. The Protarthropoda where there was one pre-oral segment, *Peripatus* belonging here. Finally, the Eurarthropoda, containing all other arthropods.

3. Schemes such as those of Snodgrass, where the ancestral annelid type from which arthropods were evolved is called the Lobopoda. This leads to the Protarthropoda, with full arthropod design, *Peripatus* being of recent descent from these animals. The Protarthropoda gives rise to the Protrilobita leading to the modern Trilobita and Chelicerata and another line from them, the Promandibulata gives the Mandibulata which includes the Crustacea, Myriapoda and Insecta.

These three schemes represent only a fraction of the classificatory systems proposed for Arthropoda. It may seem surprising that no definite scheme acceptable to all authorities has yet been evolved but arthropods show again and again that convergence of structure in the groups produces remarkable similarities. For example, the mandible, so similar in Crustacea and Insecta, is nevertheless evolved through quite different series, being the base of the limb in the former and the whole limb in the latter. The compound eye is almost identical in structure yet of independent origin in the Myriapoda, the Insecta and the Crustacea and finally, the tracheal system has evolved independently in Insecta and Arachnida and in the Crustacea. No system of classification can be produced which does not imply these remarkable convergences and systems which utilize these similarities are not expressing the natural relationships of the animals within the Arthropoda.

In the following Table (Table 1) is given the classification of Arthropoda which seems to reflect the modern idea of the relationships within the group. While the cleft between the Onychophora Myriapoda, Insecta line and that of the others is deep, they may have been closely related in the primitive unknown fauna of Pre-Cambrian times from which they must have originated.

Table 1   Phylum Arthropoda

Class	Sub-class	Order
Prot-onychophora		
Onychophora		
Pauropoda		
Myriapoda	Diplopoda	Polyxenida
		Glomerida
		Glomeridesmida
		Polydesmida
		Julida
		Spirobolida
		Spirostreptida
		Cambalida
	Chilopoda	Geophilomorpha
		Scolopendromorpha
		Lithobiomorpha
		Scutigeromorpha
Symphyla		
Insecta	Apterygota	Collembola
		Protura
		Diplura
		Monura
		Thysanura
	Pterygota	

Sub-class	Division	Order
Pterygota	Palaeoptera	Eupalaeodictyoptera
		Megasecoptera
		Ephemeroptera
		Odonata
	Polyneoptera	Dictyoptera
		Isoptera
		Plecoptera
		Notoptera
		Phasmoptera
		Orthoptera
		Dermaptera
	Oligoneoptera	Coleoptera
		Megaloptera
		Neuroptera
		Mecoptera
		Trichoptera
		Lepidoptera
		Diptera
		Siphonaptera
		Hymenoptera

Table 1 Phylum Arthropoda (continued)

Sub-class	Division	Order
Pterygota (cont)	Paraneoptera	Psocoptera Mallophaga Anoplura Thysanoptera Hemiptera

Where possible all insect orders are given the ending -ptera.

Class	Sub-class	Order
Trilobita		Protoparia
		Hypoparia
		Opisthoparia
		Proparia
Merostomoidea	Prochelicerata	Limulava
		Leanchoilida
	Emeraldellida	Emeraldella
		Naraoidea
	Cheloniellida	Cheloniellonida
Marrellomorpha		Marrellina
		Mimetasterida
Pseudocrustacea		Burgessida
		Waptida
Merostomata	Xiphosura	Limulida
		Synziphosura
		Aglaspida
	Eurypterida	
Arachnida		Scorpiones
		Pseudoscorpiones
		Thelyphonida
		Schizopeltida
		Amblypygi
		Palpigrada
		Ricinulei
		Solifugae
		Opiliones
		Araneae
		Acari

Table 1   Phylum Arthropoda (continued)

Class	Sub-class	Super-order	Order
Crustacea	Branchiopoda		Anostraca
			Notostraca
			Conchostraca
			Cladocera
			Cephalocarida
	Ostracoda		Myodocopa
			Cladocopa
			Podocopa
			Platycopa
	Copepoda		Calanoida
			Monstrilloida
			Cyclopoida
			Harpacticoida
			Notodelphyoida
			Caligoida
			Lernaeopodoida
	Mystacocarida		Derocheilocarida
	Branchiura		
	Cirripedia		Thoracica
			Acrothoracica
			Rhizocephala
			Ascothoracica
	Malacostraca	Leptostraca	Nebaliacea
		Syncarida	Anaspidacea
			Bathynellacea
		Pancarida	Thermosbaenacea
		Peracarida	Mysidacea
			Cumacea
			Tanaidacea
			Gnathiidea
			Isopoda
			Spelaeogriphacea
			Amphipoda
		Hoplocarida	Stomatopoda
		Eucarida	Euphausiacea
			Decapoda

The Decapoda are sometimes sub-divided to give the Macura (lobsters, shrimps), the Anomura (Hermit crabs) and the Brachyura (crabs), but now more often into the Natantia (prawns and shrimps) and the Reptantia (crawfish, lobsters and crabs), it being recognized that the crabs are the end products of several lines of evolution.

10

Adjustment to Ways of Life

INTRODUCTION

The features of arthropod design considered here are those which adapt it in detail to one of the many diverse environments available to it and/or to penetrate into a major habitat previously closed to it. In an animal as complex as an arthropod there are many different features at this level which can adjust the arthropod to one way of life rather than another. The number of adaptations found in any one case is a good indication of the degree of commitment of the animal to that way of life. Commitment to one way of life does not prevent further change enabling the animal to live in another way; the manner in which a change can occur will depend upon the properties of the starting population and the nature of the natural selective forces operating upon it. Many of the alternative features are qualitatively rather than quantitatively different.

An approach to this aspect of arthropod design is facilitated by concentrating upon those features in each of the sub-programmes that confers upon the animal some property which, depending upon the contents of the other sub-programmes, helps to adapt the animal to a number of different ways of life. Specialization in fat metabolism in the C.B.C. sub-programme can be an adaptive feature for: a predominantly liquid diet, floatation in water, migratory flight, long periods of fasting, or water production in dry conditions, depending upon the contents of the other sub-programmes. By considering the different combinations of the data so presented, a wide but not exhaustive range of adaptations for different ways of life can be described. This range can be extended by enlarging the contents of each sub-programme. Some such set of adaptations must be present in every individual arthropod.

CHANGES IN THE C.B.C. SUB-PROGRAMME

Since by definition this sub-programme contains information necessary to set up the cellular properties and chemical reactions found in arthropods which are also widespread in other animal Phyla, its changes will not be peculiar to them. Nevertheless they form an important part of and sometimes all the adaptive commitment displayed by an arthropod. The ways in which the contents of this sub-programme can change are considered under three headings: an emphasis on one or a few of several properties all of which are present in the animal; the loss of some of these properties and the retention of others; the generation of new properties.

Adaptive change by emphasis on some metabolic pathways in preference to others is illustrated by the mechanisms of energy production. The production of energy in the form of phosphate bonds from the metabolism of carbohydrates, fats and proteins involves a web of chemical relationships, some of which are illustrated in the diagram. However, even when all

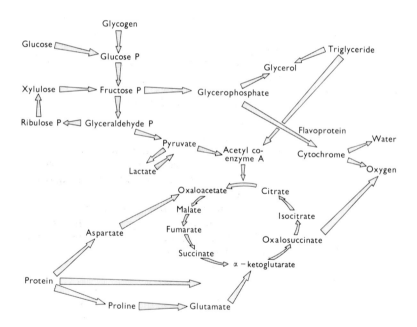

the enzymes are present in detectable amounts it does not follow that all the pathways are extensively used by the animal.

The appropriate pathway will be that dominant one where the animal has either carbohydrate, fat or protein as its main original source of material. Modifications may occur where there has been a change of diet. In some insects, where the flight muscles are geared to fat metabolism and the animal is now a predominantly carbohydrate eater, the fat metabolism persists and the carbohydrate has to be converted to fat before use. Fat may also dominate over carbohydrate where large reserves of fat have to be stored, e.g. the food reserve to supply energy on the migration flights of locusts and butterflies.

The pentose pathway for energy production from glucose does not require intermediate cycles in the degradation of glucose to carbon dioxide and water. It is likely to be emphasized where either ribose sugars and/or pyridine nucleotides are required for use in fatty acid metabolism.

Glycolysis leads to the formation of pyruvate and glycerophosphate. In the presence of sufficient oxygen, pyruvate feeds quickly into the tricarboxylic acid cycle but, when oxygen is in short supply, pyruvate is converted into lactate and lactic acid and held here until it can revert and then enter the cycle. Oxygen shortage is rare in terrestrial arthropods possessing tracheal systems; lactic dehydrogenase and oxygen debt is uncommon in these animals.

α-glycerophosphate may come from glycolysis, as mentioned above, or be formed from glycerol produced by the degradation of triglycerides in fat metabolism. It is metabolized via the α-glycerophosphate cycle which leads directly to the flavoproteins and cytochromes, not to the tricarboxylic acid cycle. It seems to be emphasized where there is a very high energy turnover rate such as occurs in the flight muscle of insects. This is an extremely 'fast' pathway and perhaps accounts for the high oxygen consumption found in these animals.

Glycerol may be formed from α-glycerophosphate by the action of a phosphomonoesterase and from dihydroxyacetone by a polyol dehydrogenase. These reactions are associated with anaerobic respiration and may serve an adaptive function in hibernating arthropods where the accumulation of glycerol acts as an antifreeze, preventing ice damage and osmotic stress to the tissues when they are exposed to very low temperatures.

The tricarboxylic acid cycle may be universally present. It can accept pyruvate and fatty acids through the medium of acetyl co-enzyme A, and degrade them to carbon dioxide and hydrogen. This cycle can also accept amino acids such as glutamine and proline and this energy source is emphasized in protein feeding animals such as tsetse flies. The hydrogen ions produced by this cycle couple with oxygen to produce water, mainly through the action of a series of cytochromes. For a given weight of metabolite most water is produced via this means from fat and this aspect is likely to be emphasized where there is need for the animal to produce its own water.

The energy released by these reactions is trapped in phosphate bonds as adenosine mono-, di- or tri- phosphates. The enzymes which release this energy to do work are the apyrases. In insects these enzymes have very high activation energies and hence work better at high environmental temperatures. This is perhaps one reason why insects are so temperature sensitive and favour relatively high temperatures. Little is known about apyrase properties in Crustacea, where low activation energies such as occur in other groups of animals (fish) would be more suitable to the cool aquatic environments which they inhabit.

In many arthropods the loss of part of the biochemical web has restricted the animals to a more limited set of responses, this loss often being associated with, and therefore said to form, part of the animal's adaptation to, some particular way of life. The loss is compensated by a deeper commitment of the other sub-programmes to this way of life, or by the saving of energy now no longer needed to produce rarely or never used enzyme systems. Insects, unlike other arthropods, are unable to synthesize their full requirements of choline and must rely on their diet for supply of this substance. *Drosophila* has flight muscle metabolism closely linked with carbohydrate usage and cannot use fat. Flight is impossible once its glycogen stores are exhausted although plenty of fat may be present. In *Culex* mosquitoes, once the carbohydrate has been exhausted the animals cannot fly for a period until some of the fat has been converted into carbohydrate. In *Glossina*, energy for flight is obtained from the amino acid proline.

Arthropods lack the ciliary and mucus cells found in nearly all other animal groups and thus cannot form the mucus—ciliary food collection mechanisms which are so well developed in other groups.

The appearance of new features at this level is confined to the development in a species of some characteristic not found in its nearest relatives, although it may occur elsewhere in the Phylum or in the animal kingdom. For example, haemoglobin appears sporadically throughout the arthropods, usually in aquatic animals likely to encounter very low oxygen tensions. Although the haemoglobin differs in detail from the haemoglobin of other animals, it clearly belongs to this group of chemical substances and is a member of the C.B.C. sub-programme.

The substance chitin and the protein 'arthropodin' which form an important part of the arthropod cuticle are found in many other Phyla. In some insect and crustacean cuticles where high elasticity is required a special protein resilin is produced and is an adaptation to this requirement. Resilin is not known outside the arthropods and it is possible that it could only have been produced in association with a cuticle organized upon arthropodan lines. However, previous experience makes one doubt whether any biochemical product is exclusively confined to one Phylum.

CHANGES IN THE G.A.D. SUB-PROGRAMME

Every anatomical feature of the arthropod body has or had a function. Each anatomical feature is therefore an adaptation enabling the animal possessing it to do something which, in its absence, would be difficult or impossible to accomplish. The adaptive significance of the anatomy of the arthropod can be considered at three levels:

1. Those which are implicit in the design of the arthropod body at Class level as, for example, the anatomical features which dictate the terrestrial and locomotory features of the Myriapoda or Insecta, or the locomotory, feeding and aquatic nature of the Crustacea. This aspect of arthropod design has been dealt with in the previous chapters.

2. Within each of these designs there have been changes in anatomy which have led to improvements of these basic designs, such as changes in body form and limb design which have led to more efficient locomotion, whether walking or swimming; more efficient vision, whether by ocelli, as in spiders, or compound eye development in other arthropods. Developments occur too which increase the capacity of the organs to do work, as in tissue design in guts. These have been dealt with in previous chapters.

3. Finally, there are changes in the anatomy of each of these basic plans which enable the animal to live in a manner different from that conferred by the original design properties; for example, the invasion of dry land by Crustacea or water by Insecta.

It is this last set of design modifications that is now to be considered, those which adapt the animal to live in one way rather than another. The relative nature of the adaptations is clearly indicated in the above paragraphs, and the adaptive radiation of the numerous arthropod families and genera provides data which make this sub-programme one of the most extensive in the arthropod programme.

For this reason only the major changes can be given here and in general terms; that is, those features of the arthropod body which tend to appear again and again in different arthropod groups and whose presence indicate that the animal is committed to a definite mode of life.

Changes in body form

In many groups of arthropods the body of some species is longer and slimmer than is judged normal for the group (Fig. 10.1). In the Chilopoda an elongate body is associated with the animal's ability to penetrate loose soil or vegetable debris, where a thin flexible form of many segments and many pairs of appendages aid manœuvring under these conditions. In the Amphipoda the elongate body form of *Caprella* is associated with disguise amongst the slender fronds of the seaweed which the animal inhabits. In

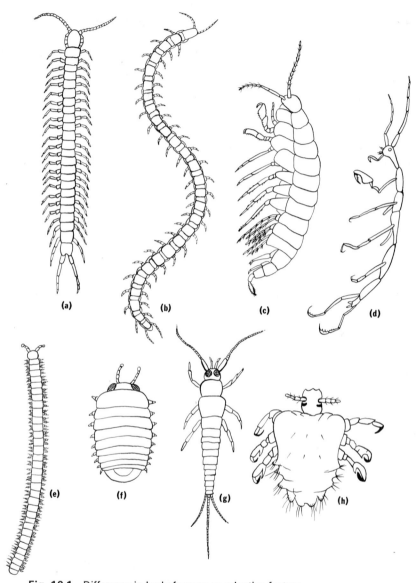

Fig. 10.1 Difference in body form as an adaptive feature.
Lengthening of body form, Myriapoda, Chilopoda: (**a**) General body form. (**b**) *Geophilus* lengthening associated with locomotion through the soil; lengthening in Crustacea, Amphipoda: (**c**) *Gammarus*, typical body form. (**d**) *Caprella*, living amongst sea-weeds.
Shortening of body form, Myriapoda, Diplopoda: (**e**) *Julus*, typical body form. (**f**) *Glomeris*, shortened form probably associated with rolling up as a defence mechanism; shortening, Insecta: (**g**) *Machilis*, typical active form. (**h**) *Phthirus* (crab louse), shortened form associated with ecto-parasitic existence.

parasitic forms such as the mite, *Demodex folliculorum*, it is associated with living in long tubular habitatats, in this case the hair follicles of mammals.

Shortening of the body form (Fig. 10.1), reduces flexibility of the lateral movements of the body and often aids the animal in sideways movement, i.e. body lice (Anoplura), in insects, or the whale louse (*Cyamus*), in

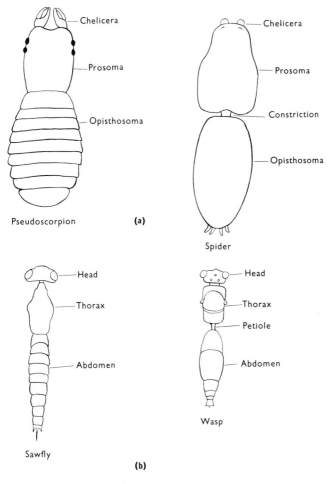

Fig. 10.2   Increase in mobility of body movement due to the constriction of one or more body segments. **(a)** Arachnida, pseudoscorpion showing normal body form compared with a spider with the body constricted between the prosoma & opisthosoma. **(b)** Insecta, Hymenoptera, a sawfly showing normal body form and a wasp *Tiphia* with the body constricted at the first abdominal segment.

Crustacea. It is often associated with fast running and can enable the animal to roll up into a ball as a defence mechanism (*Glomeris* in millipedes, woodlice in Crustacea).

An increase in flexibility between one part of the body and another is often accomplished by restriction of the diameter of the body either between two segments or within a segment, with the development of appropriate articulations (Fig. 10.2). This development is shown by many arachnids such as spiders and the solifugids, where the opisthosoma is freely movable relative to the prosoma. Many of the Hymenoptera (ants, bees, wasps and ichneumon flies) have a similar restriction between the first and second abdominal segments which gives mobility to the abdomen relative to the thorax and greatly aids the complex processes of egg laying and the stinging of prey and of enemies. In muscoid Diptera the constriction is between the last thoracic and first abdominal segments.

An asymmetrical body form, that is one in which the left and right sides are different, occurs in the Anomura (decapod Crustacea, Fig. 10.3). In the hermit crabs the abdomen is soft and is protected from damage and predators by the animal fitting it into an empty snail shell. To fit the spiral cavity of the shell and anchor the animal to it, the right side of the abdomen is shorter than the left and lacks appendages; the tip of the abdomen has hook-like appendages for fastening to the tip of the shell; those of the left side are notably larger than those of the right. In the land crab, *Birgus*, which is related to these animals, and the stone crab, *Lithodes*, the animals no longer adopt this habit and the abdomen is relatively small and covered with hard sclerotized plates. The asymmetry of the abdomen, however, still persists.

In the barnacles (Cirripedia, Crustacea), the animal during the course of development attaches to the substratum by an adhesive gland situated on the head. The body shape (Fig. 10.3) is transformed to one very different in appearance from the typical crustacean form, but well adapted to utilizing the setae-covered thoracic limbs for food gathering while maintaining a sedentary existence.

Often the need for maintaining the typical arthropod body form is no longer important and the shape of the animal is dictated by the habitat which it occupies. The final shape may be so different from that normally associated with arthropods that, but for the developmental stages of the animal, it would hardly be recognized as belonging to this Phylum (Fig. 10.3). This is most frequently seen in parasitic forms, especially in the Copepoda and Isopoda (Crustacea). In the parasitic copepod *Lernaeocera branchialis*, the body form of the adult female which lives on the gills of the whiting is that of a sac covered with a thin cuticle, having a much-branched process at the anterior end, and two egg tubes arising from about two-thirds the way down the body. In *Xenocoeloma* sp. from the body cavity

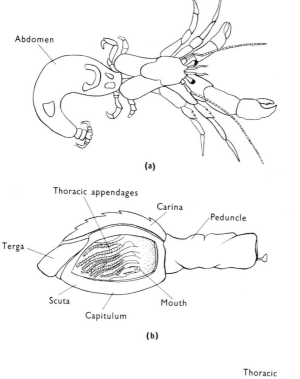

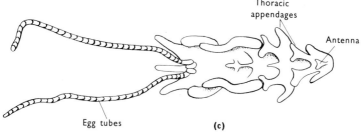

Fig. 10.3   Unusual body forms. **(a)** *Eupagurus* sp. (Crustacea, Decapoda). **(b)** *Lepas* sp. (Crustacea, Cirripedia). **(c)** *Chondracanthus* (Crustacea, Copepoda).

of the annelid worm *Polycirrus* sp. the animal is a mass of tissue, lacking appendages and even a cuticle. Such distorted and atypical forms are a mark of an animal deeply committed to parasitic existence and are often the end products of evolutionary lines that lead from the normal through different degrees of modification to this end result.

Changes in appendages

Much has already been said about the structure and function of appendages in a previous chapter and a great deal about the habits of the animal can be deduced from its appendage pattern.

In animals with an active body form, overall elongation of the appendages is associated with increased running speed (Fig. 10.4). This may be seen in the centipede, *Scutigera*, or the crabs of the family Ocypodidae.

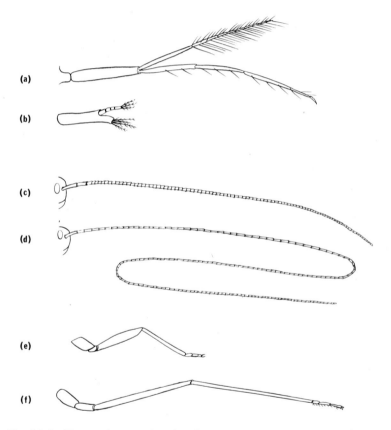

(a)

(b)

(c)

(d)

(e)

(f)

Fig. 10.4 Changes in appendage length.
Crustacean limbs: (**a**) appendage in phyllosoma larvae of *Palinurus* elongate to aid floatation. (**b**) Normal appendage from *Lucifer*.
Insect antennae: (**c**) normal antennae of longhorned grasshopper *Metrioptera* sp. (**d**) Elongate antennae of *Dolichopoda* adapted to cave life.
Insect legs: (**e**) normal leg of a hemipteran *Gastrodes*. (**f**) Elongate leg from the hemipteran *Hydrometra* adapted for movement on pond surfaces.

Long legs often occur in arthropods where the body form is less well suited to rapid locomotion. In pond skaters and insects which live on the surface film of water, or in abysal pycnogonids like *Colossendus* walking on the abysal ooze where the surface on which the animal is moving is not very strong, long legs spread the weight of the animal over a wide area of the surface film which is then able to bear their weight. In many planktonic forms such as the larvae of the crab, *Palinurus*, where floatation is essential, the long appendages increase the surface area of the animal and hence the friction, thus delaying sinking. Elongation of the limbs may also occur where the animal is very dependent upon tactile information about its surroundings; for example, the phalangid *Buresiolla karamani* (Arachnida) from the caves of Bosnia has enormously elongated appendages, even for this group of animals.

Limbs shorter than usual but otherwise well developed are associated with animals that require forward thrust to penetrate through their habitat; for example, many of the millipedes amongst the Myriapoda, or the mole crickets amongst the Orthoptera. Short, weakly developed limbs are a sign of reduced locomotory activity and a semi-sedentary existence.

Limbs which tend to be short and have the terminal pre-tarsus modified to form a hook, or are very short and hook-like, indicate that the animal spends much of its time clinging to a surface from which it may well be dislodged, for example, Branchiura (Crustacea), which are external parasites on fish, and ectoparasitic insects such as lice.

Elongation of the antennae, antennules or sensory appendages independently of the legs occurs where there is dependence upon tactile information from the environment. In the Myriapoda a 'normal' *Lithobius* has antennae of moderate length made up of 20–50 segments. In *Lithobius matulici* from caves in Europe the antennae are greatly elongated and made up of 106 segments. Amongst the Orthoptera the cave-dwelling Australian *Speleotettrix tindalei* has very long antennae even in a group characterized by long ones. Many deep sea Crustacea also have antennae and antennules relatively long compared with their shallow-water-dwelling relatives.

Modification of the mouthparts from the standard type of the group so that the animal can exploit a different food source is a very common occurrence. Amongst mandibulate particulate feeders the main modifications are for fluid feeding, with or without the ability to penetrate skin or plant tissues to reach the fluid source. In Hemiptera (Insecta) (Fig. 10.5), the mouthparts are elongated, the maxillae being long and thinly pointed at the tip. Each maxilla is semi-circular in cross-section with two grooves running the full length of its inner surface. The maxillae are closely applied together in the mid-line so that a tube is formed from each pair of grooves. A locking ridge and groove running in the anterior and posterior edges of the maxillae serve to hold them closely together. The anterior tube leads to the mouth where the pharynx is modified to form a powerful sucking pump drawing fluid up the tube into the gut. The posterior tube conducts saliva from the salivary glands to the tip of the maxillae. The mandibles are elongate flattened blades lying each side of the maxillae and capable of penetrating plant and animal tissues. Both maxillae and mandibles lie in a deep groove in the anterior surface of the elongate labium and are covered anteriorly at the base by an elongate labrum. In carnivorous bugs where the mouthparts tend to be short and stout the animal simply stabs its prey, the mandibles

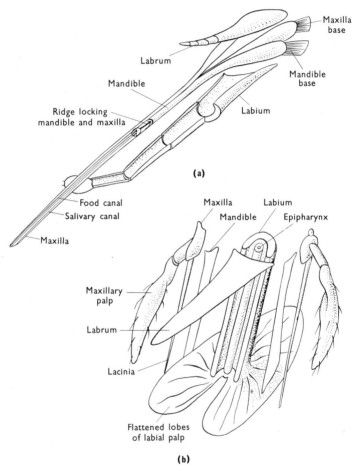

Fig. 10.5   Diagrams illustrating two modifications of the primitive biting mouth-parts of insects for taking in liquid food. (a) Hemiptera. (b) *Tabanus*.

and maxillae entering the tissues, the softer labium bending and remaining externally. In plant-feeding bugs where the mandibles and maxillae may be very long and are coiled in the head, they are inserted little by little into the tissue by the action of the labium which grasps them, pushes them a little way in, releases its hold, moves a little way back up the mandibles and maxillae, grasps and then thrusts them further in.

In mosquitoes the tube which is inserted into the tissues is formed from the labrum and the hypopharynx and is protected when not in use by the labium. The mandibles and maxillae are reduced to long delicate stylets but are still able to penetrate the tissues.

When fluid is taken from exposed places so that penetration of tissues is not necessary, the tube or proboscis is formed from some of the mouthparts, the others tending to atrophy. In most butterflies and moths a long flexible tube is formed from the maxillae in which the galeae elongate, are channelled on their inner surface and held together in the mid-line by means of hooks and interlocking spines. Mandibles are absent and the other mouthparts relatively reduced. In the muscoid Diptera (house flies, blow-flies), the labium is elongate and expanded distally to form two fleshy lobes imprinted with a system of grooves (pseudotracheae) on the surface. When applied to free fluid the liquid enters the pseudotracheae and passes to the mouth under the combined forces of capillarity and a powerful pharyngeal muscular pump. Mandibles are absent but the other mouthparts help to form the complex base of the proboscis. In the Brachycera (Diptera, horseflies) (Fig. 10.5), the mandibles and maxillae are still present and well developed, being used to cut the skin of the animal's prey, the fleshy proboscis then taking up the fluid, blood or tissue fluid, which oozes out.

In many Crustacea that have adopted a parasitic mode of life the mouth parts are similarly modified to form piercing and sucking tubes, the precise segmental appendages involved depending upon the evolutionary line being considered.

The sense organs

The degree of the development of the sense organ is an expression of the animal's dependence upon it. Enlargement of the eye above that normally found in the group indicates that vision is an important sense for those animals. In Tabanid flies the eye is enlarged to occupy almost all the dorsal and lateral sides of the head, the eyes meeting in the mid dorsal line (Fig. 10.6). These animals tend to sit on stones or twigs in broad sunlight and fly to passing insects, either recognizing them as prey or as mates. In the Arachnida, which lack compound eyes, the ocelli may also be enlarged where vision is important in prey detection as illustrated by the large dorsal ocelli of the spider, *Dinopus* sp. (Fig. 10.6). This animal generally tends to live in

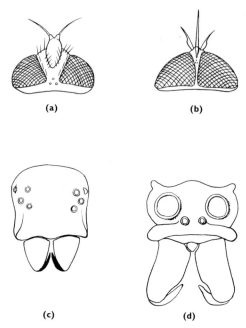

Fig. 10.6 Increased development of eyes in two groups of arthropods. (a) Tachinid (Insecta, Diptera) showing normal development of eyes within the group. (b) Tabanid (Insecta, Diptera) showing increased size of eyes in the male. (c) *Parabomis* (Arachnida, Araneae) showing 'normal' ocellar equipment for spiders. (d) *Dinopus* (Arachnida, Araneae) showing increase in size of a pair of ocelli.

well-lighted habitats, but eye enlargement may also occur as a response to visual requirements in dimly lighted places. For example, the deep sea crab *Geryon affinus* has very large eyes and some of the deep sea Ostracoda have large and complex ocellar equipment.

In dark or dimly lighted places the arthropod may also respond by placing less reliance on sight with consequent regression of its visual apparatus. The pattern of regression appears always to follow different courses in the different arthropod groups.

In the compound eyes of the Amphipoda (Crustacea) regression of the eye always starts with the loss of the crystalline cones, followed by regression of the rhabdome and loss of the retinula cells. The eye is then represented peripherally only by a thickened epidermis close to the optic lobes; finally the optic ganglia of the brain are lost. This is a progressive change, the eye of the adult animal always being further along the path of regression than that of the young. The reverse is true in Insecta where, although approximately the same pattern of regression is found, it is achieved by the eye developing very much more slowly than the rest of the body. Hence, when the animal is adult the eye will still be in a nearly embryonic state, but better developed than in the young animal.

In spiders, where the visual apparatus consists of a pattern of ocelli on the prosoma, loss of visual apparatus has occurred in many evolutionary lines. The same pattern is always followed. The median anterior ocelli are first affected, the pigment disappearing; then the ocelli atrophy and vanish: then the posterior lateral ocelli disappear, followed by the anterior lateral ocelli.

The integument

In most terrestrial arthropods, where the exoskeleton supports the tissues of the body to a greater extent than occurs in aquatic animals as well as providing the basis for muscular origins and insertions, the cuticle is thick and strong. In aquatic and parasitic animals, where some support for the body is derived from the greater density of the watery medium, the cuticle tends to be thinner. Notable exceptions occur: in many of the large decapod Crustacea a very thick cuticle is developed and may serve to support the greater bulk of the tissues, to provide adequate skeletal arrangements for the larger and more powerful muscles, to increase the density of the animal to help it to stay on the bottom, or to provide a heavy armour for protection against predators. In terrestrial animals in very protected environments, i.e. larvae of many aculeate Hymenoptera (ants, bees, wasps), or the nymphs of Isoptera (termites), the cuticle is thin and flexible. Often the muscle origins and insertions are attached to harder thicker cuticle which may stand out as darker ridges and invaginations from the rest of the cuticle.

Very often the cuticle surface is patterned with an elaborate array of ridges or pitted with invaginations. These probably serve to strengthen the cuticle due to the mechanical properties of the patterns they assume, while avoiding the weight that would be there if simple solid plates were used. The cuticle may often develop setae (aculeate Hymenoptera), or scales (Lepidoptera and Diptera), which cover the wings and body, such coatings having a heat insulating effect largely directed to keeping the metabolic heat in and the body warm under cool environmental conditions.

Changes may also occur on the cuticular surface which are associated with respiration and tend to occur in terrestrial animals, mostly insects, that have re-entered an aquatic environment. The structure of the integument changes so that a thin film of air can be held close to the body surface, even when the animal is submerged. This air film opens to a spiracle. The cuticular surface is produced into a large number of small projections set very close together. The surface is hydrophobe so that water cannot penetrate between them and wet the cuticle. The space between them always contains air. When the animal is submerged the air film so trapped is the source of oxygen supply to the animal. As oxygen is withdrawn from it the pressure within the film drops. Water cannot enter the region to keep the air there at constant pressure, but oxygen can diffuse from the water into the air space. This it can do at a rate which is adequate to meet a low oxygen consumption by the tissues almost indefinitely or to meet a high one for a short period, thus relieving the animal of coming to the surface to renew its oxygen supply at such frequent intervals as would otherwise be necessary (Fig. 10.7). Similar arrangements may be found on the egg shells

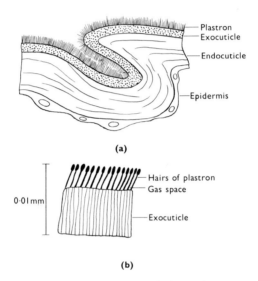

Fig. 10.7 Diagram of the plastron of *Aphelocheirus* (Insecta, Hemiptera). (a) Section through part of the ventral surface of the thorax showing variation in length of plastron hairs. (b) A few of the hairs drawn on a larger scale. (Redrawn from Thorpe & Crisp, 1947)

of many terrestrial arthropods where the eggs are in danger of submersion when in their normal environment.

Something has already been said about the colour structure in arthropods. Now we are concerned with the adaptive significance of the colours and the patterns they form. In many cases the colour of the animal matches that of its environment so that the animal is hard to see and thus it gains protection from its predators which have colour vision and hunt by sight. Thus many arthropods living on green vegetation are green; in deserts, sand coloured; on tree trunks, black and grey; or they match the colours of the flowers they frequent. Many may lack colour and are almost transparent as occurs in many planktonic Crustacea. Inhabitants of dark places, such as cave or soil dwelling arthropods, tend to be white, but many deep-sea Crustacea are red in colour.

Many arthropods are small and uniformly coloured in a way that matches only a part of their habitat, black or dull coloured animals being well hidden in the many shadows cast in any environment. Where the animal is large enough to rest over multi-coloured parts of its environment its body is often marked with elaborate multi-coloured patterns which so coincide with those of the environment that the animal is indistinguishable from them; for example, the pine hawk moth.

Many arthropods are brilliantly coloured so that instead of matching their environment they are extremely conspicuous in it. Such colours tend to be black and yellow, red, or white, and the patterns they form crude. Arthropods marked like this usually have a defence mechanism, either a sting, some obnoxious fluid which they produce, or unpleasant-tasting qualities.

This gives them protection, because once a young predator has sampled one of these insects and found it to be inedible, the bold colour pattern is remembered and other similar brightly coloured forms avoided in the future. Thus the population, rather than the individual, is protected by this means. This type of protection is increased by the animals tending to congregate in groups, to display themselves conspicuously and by different species tending to resemble each other (Mullerian mimicry), so that fewer colour patterns have to be distinguished by the predator and one sample will protect a number of species.

A number of edible animals lacking any offensive means of protection gain some relief from predation by resembling the inedible forms (Batesian mimicry).

Colour patterns are also developed in connection with mating behaviour. Such patterns are often very elaborate and composed of bright colours, but frequently situated so that they are not displayed all the time, only at the appropriate moment in courtship. Many arthropods, particularly Crustacea from the moderate depths of the sea, have luminescent organs arranged in complex patterns over their bodies. These appear to aid in sex recognition, as does the luminescence of many nocturnal terrestrial arthropods such as the fire-flies.

Changes in the gut

Much has already been said about changes in the gut which increase its efficiency but, in addition to these, some adaptations occur which need mentioning here.

1. Certain regions of the gut may be enlarged and form reservoirs for Protozoa which have a symbiotic relationship with the arthropod. In some but not all termites the hind-gut is enlarged to form an asymmetrical sac in the lumen of which there is a large fauna of Protozoa of the family Trichonymphidae. These Protozoa which are not able to live and multiply away from this environment, break down the cellulose of the termite's food to simpler carbohydrates, some of which are used by the insect.

2. Some insects take in large quantities of fluid, usually plant sap which is low in protein content, so that the gut has to process a large amount of material to get the necessary amounts of amino acids. Here the fluid is concentrated before it enters the mid-gut. The mid-gut is long and curved so that the hind-gut and Malpighian tubules come to lie close to the posterior part of the crop; fluid passing from the crop or anterior mid-gut is picked up by the hind-gut and Malpighian tubules so that only concentrated material enters the bulk of the mid-gut and the extra fluid is quickly removed from the haemocoele. In some Hemiptera the hind- and fore-gut form a complex organ, water being readily transferred from fore- to hind-gut without entering the haemocoele.

3. In many arthropods the lumen of the gut is not continuous down the length of the tube but is often blocked between the mid-gut and the hind-gut so that passage of food cannot occur. This is often found where the animal would foul its habitat by excretory matter, for example in larvae of social insects living in cells. The waste accumulates in the gut until the animal pupates when it is ejected from the gut lumen opening and removed by the attendant worker insect.

4. The gut may be rudimentary or absent. It is often rudimentary in insects like mayflies where the adult does not feed and is light for aerial flight, or in moths where the adult does not feed, or in parasitic forms, copepods, where food intake is not through a recognizable alimentary canal.

CHANGES IN THE F.R.H. SUB-PROGRAMME

Adaptations in this sub-programme independent of the changes of the G.A.D. sub-programme can be found in the races of different species of arthropods.

In the tick *Ornithodorus moubata* (Fig. 10.8), which is widely distributed in East Africa and is important as a vector of relapsing fever, investigations have shown that the population is made up of four races, each adapted to different food, temperature and humidity requirements, yet differing only in slight morphological points which are only visible when the tick is engorged with blood at its first meal. One race is abundant only in cool wet climates at high altitudes where it feeds readily on man. It occurs in greatest abundance at temperatures 20–25·6 °C and *R.H.* 86%. It is absent

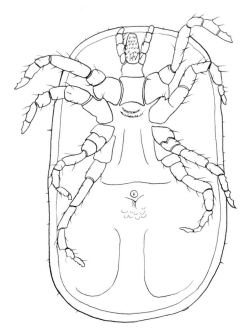

Fig. 10.8 *Ornithodorus moubata* (Arachnida, Acari) female, ventral view.

from places where the *R.H.* is consistently over 90% or less than 75%. It lays a large number of eggs. Another race prefers warm moist places and feeds mainly on domestic fowls. It is most abundant at 25·6 °C and an *R.H.* 83%. It ranges up to 28·9 °C and lays fewer eggs than the preceding form. A third race feeds mainly on wart-hogs and porcupines and lives away from human habitations. It prefers haunts where the mean temperature is 24·4 °C and the *R.H.* 72%. The fourth race is most abundant in hot arid conditions with mean temperatures of 30 °C and *R.H.* 43%.

Differences may also occur in the synchronizing signal recognition by which populations of the same species adapt to the local conditions of their environment. For example, in the large cabbage white butterfly, *Pieris brassicae*, the pupae will enter diapause according to the mean day length characteristic of the approach of winter conditions. In northern parts of its range such as Leningrad, Brest and Belgorod, a day length of 15 hours leads to diapause, representing the long summer days shortening towards autumn. In the more southern part of its range, North Caucasus, Central Asia and Abkazain, where the summer day lengths are shorter and the winter temperatures not so early in appearance, diapause is signalled by a length of 9 hours. This onset of diapause can be inhibited in southern forms where the temperature is 25–26 °C and in the northern forms where it is 29–30 °C. When diapause occurs in the northern forms at temperatures below 29 °C the day length does not vary in effectiveness with temperature, but in southern forms the effective day length is 10 hours at 18 °C and 12 hours at 23 °C. Thus while the northern form is geared to a certain winter when the animal must diapause, the southern form is more flexible and can take advantage of the occasionally warm late seasons that may occur in its habitat.

The studies by Goldschmidt of the moth, *Lymantria dispar* (Fig. 10.9), indicate that the many possible changes that can occur in the life pattern of an animal to adapt it to the requirements of its habitat behave independently of each other, so that the geographical distribution of one may not coincide with that of another.

Fig. 10.9 Female (lower figure) and male of *Lymantria dispar* L.

Lymantria dispar is widespread through Europe and Asia, including Japan, and has been introduced into America. The female moths do not fly, the animal being spread by the wind which carries young caterpillars, or by egg masses floating on bark in streams or being carried by birds. The egg has a diapause period and the young caterpillars must hatch in synchrony with the appearance of young vegetation in their habitat: growth, metamorphosis and reproduction must be completed before the autumn cold ends the life of the animal.

The diapause is shortest in the forms from Europe, W. Asia, Korea and Hokkaido, places where a long winter is followed by a short spring giving only a brief period in which vegetation is suitable for the young caterpillars to feed, so that the animal must be ready to hatch as soon as conditions are right. A long diapause period occurs in the Mediterranean forms and in Japan where a mild winter occurs; otherwise under these conditions the young would hatch on the first warm day but find no vegetation for food, hatching must be delayed until vegetation is ready. In Northern Japan where the winters are cooler there is still a long diapause period. Here the spring rise in temperature occurs much earlier than in moderate climates, but vegetation does not appear until later, hence delay again must occur in spite of temperature suitability. In Manchuria the diapause is very long.

The length of larval development varies in different geographical regions and is adapted to ensure its completion within the summer growing period. It is short in northern forms and long in southern ones. In mountain regions of Japan it is short compared with that of the immediate lowland forms. In Korea it is shorter than might be expected from the summer period; this is because in late summer the temperature is too high for the larvae to tolerate, hence growth must be complete before this occurs. Thus the distribution of this character does not coincide with that of egg diapause.

The number of larval instars is genetically determined through a series of multiple genes, giving either 4 instars in both sexes, 5 instars in both sexes, or 4 instars in one sex and 5 in the other. In Spain there are four moults in each sex; in Northern Japan four moults in the males and five in the females; in Tokyo five moults in both sexes. Elsewhere different combinations occur, but European, Korean and Hokkaido races are similar, as are the Western Japan and Turkestan races.

Goldschmidt has studied many other characters: growth rate, cell size, chromosome size, metabolic rate and larval colour inheritance. All show differences, many of which are directly adaptive or can be correlated with adaptive features. In colour, the dark forms are found in most of Europe and Asia, the light forms in South West Japan and Kyushia into Korea, and the Hokkaido forms have a unique colour pattern.

These examples show that adaptive features in a species may serve quite different environmental requirements; e.g. short larval periods relate to short summers as well as to ones ending in very high temperatures, and egg diapause is long both for mild winters and for cooler winters with sudden spring rises in temperature. These features do not necessarily coincide, therefore sub-division of the species based on one feature will produce a different result from sub-division based on another, at least in the F.R.H. sub-programme.

The amphipod crustacean *Gammarus duebeni* is widespread throughout Western Europe. Over most of its range it lives in brackish waters, but in Western Britain it lives in freshwater streams and in Ireland it is common in freshwater loughs. Comparison between animals from these two habitats shows that those living in fresh water are more able to retain the salts they can actively absorb from the environment than can the brackish-water form. The latter requires therefore a higher salt content in its environment for continued existence.

CHANGES IN THE A.R.I. SUB-PROGRAMME

This sub-programme takes account of all the programmed alternative courses of action that may occur within an individual. The individual animal is thus programmed to respond successfully to a wider range of environmental factors than would be possible if it lacked the alternative information. Before discussing this it is perhaps as well to point out that a sub-programme is a set of information, both cytoplasmic and genetic, which stems from the zygote, and whose effects when considering the organization of an arthropod make it worth while to treat it as a unit. Whilst the events of each sub-programme are subject to the laws of genetics they do not coincide with them and the alternative information referred to here can give rise to genetic or cytoplasmic determined events which confer upon the animal programmed alternative responses.

Alternatives within the programme of a single individual

One example of this has already been given, where the animal has the ability to enter the diapause state or not at a given stage in its life history according to the conditions it encounters prior to that stage being reached.

In the honey bee, *Apis mellifera*, each female larva at the moment of hatching and for the next few days has the ability to develop into a queen or worker bee; differences between the two are the larger size of the queen and the presence of well-developed ovaries and, in the worker, the presence of pollen-gathering equipment on the hind legs, wax glands on the abdomen, a relatively larger brain, a short life and a different behaviour

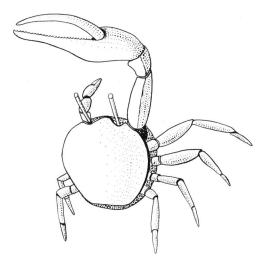

Fig. 10.10   Male *Uca pugnax.*

pattern of daily activity. Which the young larva develops into depends upon the food and care she receives from other workers during this time, great care producing a queen. In many other but not all social insects where distinct differences between worker and queen and between soldier and worker occur, each individual has the ability to develop into any of these types; which it actually becomes depends upon the treatment of the larva by other workers during its growth and development.

The structural differences between locusts, form *gregaria* when reared in a crowd, compared with form *solitaria* when reared singly, are also environmentally determined alternative responses.

Alternative behaviour patterns

In arthropods behaviour may be defined as a pattern of activity originating in responses to a requirement, enduring for a period and ending when the requirement has been attained. For example in the mud crab, *Uca* sp. (Fig. 10.10), the males burrow in the mud and will fight intrusion by any other males into their territory. The behaviour pattern starts when one male invades the territory of another. The opponents approach each other walking stiffly and hit each other alternately with their large chelae. The claws lock together and each tries to push away or overturn the other. Before this point is reached the meeker usually disengages, runs back to its own burrow and blocks the entrance with its chelae. The fighting pattern ends with this separation of the two combatants and is repeated should the occasion again arise. Some variations may occur: the loser may plug the burrow with mud and a crab that has lost its chelae will do the same. Such a crippled animal cannot fight and may be repeatedly chased into its burrow by neighbouring crabs. In another crab, *Dotilla myctiroides*, the behaviour pattern ends with the winner jumping up and down and drumming its claws on the ground.

Such patterns of activity fill the life of the animal. Feeding behaviour, defence behaviour, shelter, taxes, mating behaviour, hibernating, migrations, are all subject to patterns usually fairly constant for all members of a species, often very different between related species.

Behaviour patterns are included in the A.R.I. sub-programme because of the intimate association that exists between behaviour and body design. Although the purposes of behaviour patterns such as mating, feeding and defence are universal, the manner in which they are accomplished, the structures which bring them about, and the limitations of variation they are subject to are all products of design. The close relationship between the programme of the animal and its behaviour patterns is well illustrated by a parasitic wasp, *Habrobracon*. The wasp exhibits a common sexual difference in behaviour, the males courting the females and the female eventually stinging and laying her eggs in the larvae of a moth, *Ephestia* sp. This difference depends upon the chromosomal sex of the brain cells, not upon the gonads. Sexual mosaics occur as an abnormal development with female brain cells and testes, or male brain cells and ovaries. The animal behaves according to the properties of the brain cells, and will try to sting *Ephestia* larvae and court *Habrobracon* females respectively.

This suggests a very machine-like relationship between the design of the animal and what it does and how it behaves. This is perhaps a useful basic concept on which to place the many modifications that have produced less rigid relationships and in doing so made the animal much more adaptable to its surroundings.

First of all, to what extent must a behaviour pattern once started be completed? This depends upon the way the pattern is constructed and related to the environment.

The behaviour pattern may be built up of a sequence of stimuli and responses as, for example, in prey capture by the common spider, *Araneus diadematus*. Prey coming into contact with the web causes vibration of its threads; the spider, detecting these, moves rapidly towards the centre of disturbance. The struggling of the prey to free itself causes the spider to bite it, the biting lasting an appreciable time; this produces contact with the prey to which the spider responds by wrapping it in silk; the silk-wrapped bundle stimulates a short bite resulting in a chemical stimulus (possibly because some feeding occurs), to which the spider responds by carrying off the wrapped-up prey back to the edge of the web. This behaviour pattern depends upon stimuli (releasers) occurring in a definite sequence for its maintenance and in general the releaser for one step is the successful end of the previous one. If the end of a step is not successful the pattern ends or very frequently it is replaced by another pattern which is often quite unrelated to the activities of the original one. In many arthropods grooming and cleaning of the limbs by the mouthparts occurs and is maintained for a short period after which the animal returns to the state it was in before the original pattern started.

Behaviour patterns which happen only once in the life of the animal, such as selection of a site for copulation, or mating behaviour when only a single mating occurs, cannot be modified by experience since failure cannot be recovered. The drive for such patterns originates in the animal and may be very strong. For example in wood-boring insects the pupa is formed in a burrow separated from the outside by a slender wood partition that the adult breaks through when it hatches. Some wood wasps, can, where the timber is subsequently covered by lead, burrow through the wood and an appreciable thickness of lead to reach the outside. If, however, a paper cone is fixed outside the lead so that when the insect comes out it is enclosed in a space, it does not break through the paper cone but remains trapped until it dies. The same thickness of paper glued to the lead is easily penetrated. The drive to get out ends once the confines of the burrow are gone and the burrowing pattern is not again invoked although the animal dies under conditions it certainly has the equipment to break through.

Repetitive behaviour patterns are capable of modification and can be stopped or inhibited as a response to a particular stimulus if experimental conditions are so arranged that response does not result in the normal expected successful end. This modification takes time: often many unfavourable repetitions are necessary and disappearance may come through diminutions of the pattern rather than through its sudden extinction. Recovery occurs after an interval, the pattern usually appearing slowly. Recovery can be prevented by repeating the unfavourable situation at intervals. Alteration in the time constants of the parts of the curve gives situations very akin to learning.

In arthropods learning is usually manifested as the selection of one behaviour pattern in preference to another when the latter has by experience been found not to produce a successful end result.

These examples serve to illustrate the kind of properties that every individual arthropod must have to adjust to some particular way of life and our knowledge of the organization of the animal would be incomplete without taking them into account.

To go further than this, to deduce the way of life of an arthropod solely from a knowledge of its structure and function, or to deduce the structure and function of an arthropod solely from a knowledge of the properties of its habitat, offers many difficulties. For example, abyssal Crustacea living in a permanently lightless habitat may have well-developed eyes, normal eyes, or be eyeless. In 450 species of Brachyura (crabs) from this zone, 380 had normal eyes, 65 small eyes, and only 5, or 1% were eyeless. It may be that with increasing knowledge the difficulties illustrated by this example will be resolved, but at the moment little more can be done than to note the properties of the animal and relate them to the observed lives they lead.

FLIGHT IN INSECTS

In arthropods only the insects have developed the ability to fly. Among the wingless Apterygota, the Monura (Fig. 10.12), from the Devonian Period and the present-day Thysanura contain species closely allied to the winged Pterygota (Fig. 10.11). The earliest Pterygota from the Carboniferous Period already have well-developed wings similar to those of present-day cockroaches. The evolution of mechanisms of flight is poorly known. Nevertheless an evolutionary approach offers the best way of surveying this adaptation.

The Thysanura show many adaptations for living on the damp vertical faces of cliffs, stone walls or tree trunks. The position of the limbs on the thorax, their reduction to styli on the abdomen, the presence of ventral water-absorbing organs and the insect's method of feeding all suit this kind of habitat. The long antennae, cercal and medial filaments provide the animal with a wide sensory field of touch helpful in the detection of the approach of predators (spiders).

In such a habitat the insect can escape from predators by running, seeking shelter in crevices, or by sudden flexing of the abdomen and using the long medial filament as a lever to throw itself away from the cliff face and fall to the ground, as is commonly seen in *Machilis maritimus*. The long filaments and the legs increase air resistance so that the insect floats relatively slowly to the ground. The filaments trailing posteriorly from the animal will also ensure that it falls on its feet and can thus seek shelter immediately it touches the ground. Emphasis on this type of escape mechanism could lead to flight. It should be noted that simple jumping does not lead to escape from the predator once and for all. Spiders can also use their limbs to minimize their rate of fall, or by trailing a silk thread, float down to earth.

In an arthropod constructed like an insect, the thorax is the heaviest part of the animal and its centre of gravity. (The large ovaries of mature females are a later evolutionary development.) So that slowing down of the rate of fall, once the animal's size has exceeded the capacity of the air resistance of the trailing filaments, can best be accomplished by the development

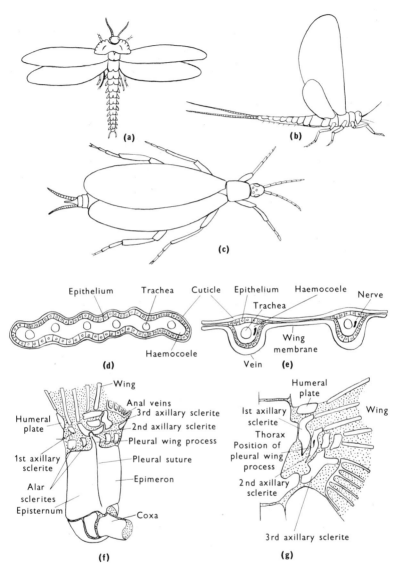

Fig. 10.11 Wings of insects. (**a**) (Paleaodictyoptera) a primitive winged insect in which the wings are normally held in an extended position. (**b**) *Ephemera* (Pterygota, Ephemeroptera) in which the wings when at rest are held vertically over the back. (**c**) *Dieconeura* sp. (Protorthoptera) U. Carboniferous. (**d**) Transverse section of a developing wing. (**e**) Transverse section of part of the same wing when fully extended in the adult. (**f**) Sclerites on the pleural and base of the wing in a primitive winged insect. (**g**) Wing base seen from the dorsal surface.

of laterally extended flaps from the dorso-lateral wall of these thoracic segments. Such flaps may be found in the larval forms of the early Pterygota (Fig. 10.12), the ones on the meso- and metathorax being much larger than those on the prothorax.

In the early instars, the air resistance of the unmodified body would be sufficient. However, as the insect gets heavier, the flaps need to appear and increase in size. During growth, the surface area increases as the square of the linear dimensions whilst the weight increases as the cube. Hence, to buoy up the heavier insect, the area of the flaps would have to get relatively larger in proportion to the rest of the body (Fig. 10.12). In the arthropod this happens in a step-like fashion. The increase at each step must have the capacity to deal with the greatest weight the insect will have in that instar. In adults, where the weight of the reproductive organs must be accommodated, these flaps will be much larger than in preceding instars.

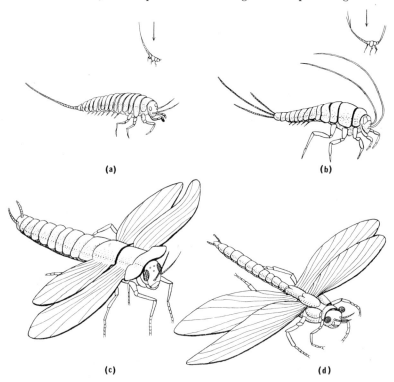

Fig. 10.12   Stages in the evolution of wings in insects. (**a**) Monura (Apterygota). Above : probable position when falling. (**b**) *Machilis* (Apterygota). (**c**) *Stenodictya* (Palaeodictyoptera, Pterygota). (**d**) *Meganeura* (Meganisoptera, Pterygota).

Large fixed projections in this position, while instantly ready for gliding, would be a hindrance to the animal in its normal movements. At an early stage a hinge develops between the base of the flap and the body wall so that the flap (wing) can be folded vertically over the thorax (Fig. 10.11). With a large wing to minimize the increase in weight, the upper and lower integuments of the wing come together so as to obliterate the haemocoelic space. Eventuaĺly the epidermis secreting the wing cuticle disappears and the wing membrane is formed by the two cuticular surfaces in close juxtaposition with each other (Fig. 10.11). Not all the haemocoelic spaces are obliterated: a pattern of channels (wing veins) is left through which blood flows, usually distally through the anterior and proximally through the posterior ones. The many diverse patterns of venation that exist in the wings of present-day insects can all be traced to a common original pattern.

The position of the development of the wings is amongst the origins of the leg muscles from the lateral tergal wall. Some of these muscles (direct flight muscles) come to be associated with the base of the wings and have a dual function of moving the wing and the leg.

When the wings are held horizontally and the animal is gliding most of its weight will be borne by these muscles. In many of the large early pterygote insects these muscles were aided by a rigid flap on the prothorax, beneath which the base of the mesothoracic wing (forewing) could fit. The anterior edge of the metathoracic wings (hindwings) fits under the posterior edge of the forewings, so that when the wings are extended horizontally they are mechanically prevented from folding vertically. This arrangement would allow the insect to glide down with the minimum of expenditure of muscular energy and would enable it to remain airborne much longer than would otherwise be possible. With the evolution of more powerful muscles and of flapping flight such an arrangement would be a hindrance and it disappears in later evolution.

Some control of the direction and steepness of the insect's glide path is possible if the wings can be moved to vary the flow of air over their surfaces. This leads to the development of an elaborate articulation of the wing to the body. Basically the wing becomes pivoted about a point on its ventral surface midway between its anterior and posterior edges, so that it is movable in all directions.

To strengthen the thorax to meet the increased stress of gliding flight the pleural wall of the wing-bearing segments becomes a strong sclerotized plate fused to the sternal plates anterior and posterior to the coxae and to the tergal plates anterior and posterior to the wing base (Fig. 10.11). The pleural plate has a dorsal projection, the pleural wing process, which comes in contact with the mid-point of the wing base. The pleural plate is also strengthened by a ridge on its inner face and a groove (pleural suture) on its outer one extending from the coxae to the tip of the pleural wing

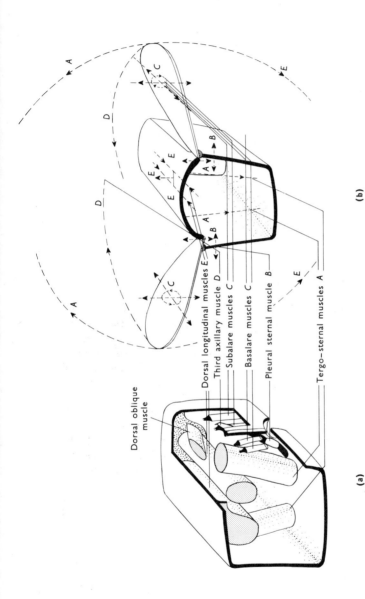

(a)

Dorsal oblique muscle

Dorsal longitudinal muscles E
Third axillary muscle D
Subalare muscles C
Basalare muscles C
Pleural sternal muscle B

Tergo−sternal muscles A

(b)

Fig. 10.13 The relationships between muscle and wing movement in insects. (a) Diagram of the flight muscles of a thoracic segment, the basalare and subalare muscles are cut short to show the pleurosternal muscles, they should continue ventrally and be inserted on the sternum and coxa. (b) Illustrates the movements of the thorax and wings by dotted lines, the arrows pointing the direction of the movement. The letter A, B, etc. associated with each muscle name is noted alongside the thoracic sclerotic movement and the wing movement, indicating the group movements for which the appropriate muscles are responsible.

process. There are three major sclerites in the wing base, the first (anterior), second and third (posterior) axillary sclerites (see Fig. 10.11). The wing base is so folded that the ventral surface of the second axillary sclerite rests on top of the pleural wing process where it is firmly held in position.

On the dorsal side, the wing is articulated to the tergum, the anterior notal process to the first and the posterior notal process to the third axillary sclerite. These articulations allow the wing to move dorsally and ventrally, anteriorly and posteriorly about its pivot point and to rotate about its horizontal axis (Fig. 10.13).

The development of flapping rather than gliding flight depends upon the insect's ability to move the wings with sufficient power and in a manner

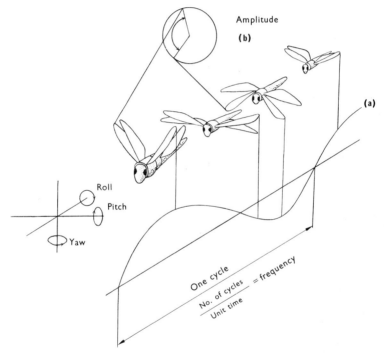

Fig. 10.14 Movement during flight in insects. Four positions assumed by the wings during a single cycle of their movement are shown. (a) is a line representing the path of the tip of a wing during flight. Vertical lines are drawn to show the relationship between the wing tip position and this line for the four stages of the cycle shown. (b) Indicates the measurement of the amplitude of the wing beat. This is the number of degrees in an arc of the circle formed by maximum position assumed by the wing at the ends of its beat. Lines are drawn from the wing tip in these positions to the chords limiting the arc. (c) Indicates the unstable movements of the body during flight.

which will push air backwards and downwards so that the animal is lifted and thrust forward.

The movements of the wings during a single cycle (Fig. 10.14) are as follows: at the commencement of the down stroke the wings are held vertically over the back of the animal; the wing then moves ventrally and anteriorly, the leading edge of the wing cutting into the air ahead of the posterior edge. This movement thrusts a volume of air downwards and backwards from the animal, lifting and giving it forward momentum. During the up stroke the wings move back to their original position, the leading edge of the wing again preceding the posterior edge. This movement causes air to be forced backwards, contributing to the forward thrust of the animal, but very little if at all to its lift. The angle at which the leading edge of the wing cuts the air stream during the down stroke affects the direction of the air thrust, the greater the angle the greater the lift. In large insects like dragonflies and hawk moths, each cycle lasts between 30–60 msecs and the animal moves about 20–30 cm. In small insects the cycle lasts for only 2–12 msecs and the animal moves about 1 cm per stroke.

At a fairly early stage in evolution the direct flight muscles were supplemented by another muscular system, the indirect flight muscles, which eventually became the main muscles supplying power to the wing movement. The direct muscles, while still retaining something of this function in primitive Pterygota, are now mainly concerned with setting the actual path through which the wing moves, almost exclusively so in the more advanced forms.

The indirect flight muscles (Fig. 10.13) in each wing-bearing segment consist of a pair of dorsal longitudinal muscles beneath the tergum on each side of the mid-line, and a pair of tergo-sternal muscles in the anterior part of the segment, arising from the sternum and inserted into the tergum lateral to the dorsal longitudinal ones. The dorsal longitudinal flight muscles arise and are inserted on phragmata which project into the body between the prothorax and the mesothorax, the mesothorax and metathorax, and the metathorax and first abdominal terga. The phragmata are firmly attached to the terga of the meso- and metathorax, and separated by membranes from those of the prothorax and first abdominal tergites. When these muscles contract they pull the phragmata together and cause the terga of the meso- and metathorax to bow upwards. This movement lifts the notal processes which also raise the first and third axillary sclerites. These sclerites lie medial to the second axillary sclerite about which the wing is pivoted, so that this raising of the base of the wing causes the down stroke of the wing membrane through a simple lever action. The tergo-sternal muscles, when they contract, pull the tergum towards the sternum; the sternum, being a strongly rigidly held plate, does not move. The tergum is thus flattened, the notal process and the axillary plates pulled down, and hence the wing moved upwards. These muscles then are antagonistic, primarily causing the tergum to oscillate up and down, a movement of very small amplitude which makes for high muscular efficiency, and which is converted to the up stroke and down stroke of the wing via the wing articulating mechanism.

The efficiency of this mechanism will be greatest when all the force generated by the muscle is converted to the lift and thrust of the insect through the air. This is nearly achieved if the wing strikes the air so that air resistance to it is at a maximum, the animal being driven forward rather than the air being displaced backward. This will happen if the wing moves smartly throughout the whole of a down stroke and/or up stroke, and less efficiently if the wing moves slowly, since air resistance increases as the speed of the body moving through it increases. With slow movement some of the air 'gets' out of the way of the wing and does not contribute to the lift and thrust. Slow flapping flight needs large wings to ensure that enough air resistance is generated to lift the animal. Since also controlled gliding is only effective for insects weighing 2 grams or more, it is not surprising that the primitive Pterygota were all large insects with relatively large wings.

To ensure the maximum speed of wing movement throughout its stroke and to reduce to a minimum the time spent in slowing down at the end of a stroke and speeding up at the beginning of another, the following mechanism has developed (Fig. 10.13). Muscles arise from the sternum and are inserted on the pleural ridge just above the coxae. When they contract they pull the pleural wing processes closer towards each other in the midline. This impedes the downward movement of the tergum when the tergo-sternal muscles contract. Thus they cannot pull the tergum downwards until they have developed their full tension, when the tergum quickly snaps from one position to another. The same mechanism affects the down stroke of the wing. When the dorsal longitudinal muscles contract the tergum cannot bow upwards until their full tension is developed, when it snaps swiftly from one position to another. With this increase in efficiency of wing movement the same lift and thrust can be developed in wings of smaller size, and in many insects where this mechanism is fully developed, such as bees, wasps and flies, the wings appear disproportionally small to the size and weight of the body. This reduction in size of the wing also greatly aids the animal in its other locomotory movements and in seeking shelter in crevices.

These mechanical developments and the high wing speed at each stroke pave the way for the development of high wing beat frequencies. In large slow fliers, e.g. butterflies, the wing beat frequency is about 9 per sec. in large insects, but in faster fliers like dragonflies, hawk moths and locusts, it is from 30–70 beats/sec.; for wasps, moderate-sized flies and bees, 300–400 per second; and in many small insects it may reach 1024 beats. These high frequencies of muscular contraction appear possible only when the muscle is working near the upper limit of its temperature. Many insects cannot fly if the body is cold, and often have to warm themselves up by muscular 'trembling' before flying.

At these high wing beat frequencies the ordinary 1:1 relationship between nervous impulse and muscular contraction no longer holds, presumably because the time they take is too long to be compatible with this rate of oscillation. What happens is that the patterned nervous outflow to the muscles delivers about three impulses per second to the indirect flight muscles which are contracting at very much higher frequencies. These impulses do not cause the muscle to contract but sensitize it so that it will contract if stretched. During flight the contraction of the dorsal longitudinal muscle bows up the tergum, stretching the tergo-sternal muscles which respond by a contraction; this in turn flattens the tergite and stretches the dorsal longitudinal muscles which respond with another contraction. In this manner an oscillating system is set up in which the contraction of one set of muscles prepares the antagonistic set to follow with a contraction in the minimum possible time. When the nerve input ceases the muscles are insensitive to stretching, the oscillation stops and the wings come to rest. Since the indirect muscles are insensitive to neural stimulation, flight cannot be started by a neural command directly to them. Alongside the dorsal longitudinal muscles are a pair of dorsal oblique muscles. These respond to direct neural stimulation and their action results in the depression of the tergum and the stretching of the dorsal longitudinal muscles which, now under the influence of the sensitizing neural outflow, respond by contraction.

The very high frequencies of muscular contraction needed for flight are associated with the development of fibrillar muscle (Chapter 1) in which the biochemical pathways for high rates of energy production are emphasized (page 194). Other systems are also modified to meet the requirements of such an active metabolic tissue. The diffusion pathway from the tracheole to the site of oxygen usage in the cell is reduced by the tracheole actually penetrating into the substance of the muscle. An adequate supply of oxygen to the tracheoles is assured by an increase in the number and diameter of the tracheae leading to the flight muscles, and by an increase in the number of tracheoles. The metabolic heat produced by the active muscle is very high and is largely dissipated by air flow through the tracheae and air sacs surrounding the muscles. The air sacs also serve to reduce heat conduction from the muscles to other tissues.

The use of fat as a fuel for flight muscle activity has already been mentioned. Since fat, when metabolized, gives more energy per gram stored than other substances, the insect, by using fat, will keep the weight factor down to a minimum. This is important in migratory insects which take long flights without feeding and are dependent upon their stored foodstuffs. The metabolism of fat results in a high rate of water production and mechanisms to facilitate water loss are developed in these insects.

The interaction of flight muscles during flight is controlled by a highly developed motor output pattern of nerve impulses. There is also developed an elaborate sensory input, monitoring the load and movement of the wings and the air flow over the body. Central nervous system integration of the two is adapted to enable the animal to alter the wing movements and thus maintain controlled flight in varying air conditions.

Other sensory information, such as visual detection of mates or predators, may initiate or terminate flight or cause the animal to change direction in flight. Thus the control of flight is linked to other central nervous patterns which are reflections of the animal's interactions with its environment.

This account, incomplete though it is, serves to illustrate the complexity of a major adaptation in arthropods; the need to know a very great deal about the properties of the initial population and the evolutionary course followed; and that a diversity of adaptive states, each able to perform the required function, is to be found.

II

The Nature, Properties and Expression of Arthropod Organization

INTRODUCTION

Studies on the comparative anatomy of the fauna of the world show that the numerous species which exist are modifications and variations upon a rather limited number of basic designs (Phyla). The members of each Phylum are linked together by being constructed upon the same basic plan. The original members of the Phyla had themselves evolved from a common ancestor and the distinct plans they now represent are survivors of more numerous and less differentiated designs. These plans are not of equal complexity: the more complex have evolved from the less complex. The arthropod plan is a complex one and the animals in which it appeared were founded upon designs which had already been subjected to a long continued evolutionary change.

The common ancestor of all animals now grouped together as arthropods may not have been an arthropod. The distinctive features of this design were perhaps independently acquired in three separate evolutionary lines. However such an ancestor must have had in its design many of the basic properties found in arthropods: the haemocoele, exoskeleton and segmentally arranged appendages, even if independently acquired, have imprinted upon their possessors a design whose uniformity contrasts greatly with the differences between them and those of other Phyla. It is therefore justifiable to consider it as a single plan quite distinct from that of all other animals, and one of the great designs that have appeared in the world's fauna.

In every case the arthropod design is expressed in an individual animal which, as a zygote, throughout its growth and development, and as an adult, exhibits a complexity of features, some widespread throughout the animal kingdom, and some of arthropodal uniqueness. Of the latter, some relate to general arthropod design, others, though expressed with due regard to this design, are concerned with the animal's ability to exploit and survive in a given environment.

Accounts have been given of the basic features of the arthropod design and its main variations found in present-day representatives of the group. Consideration will now be given to the expression of the total properties of the individual arthropod as follows: how to think of an animal so that all its properties can be taken into account; how to set out these properties so that their differences can be briefly stated; how to choose from such an array a set of properties which specifies an organism fulfilling a particular requirement and finally how to find an arthropod which has the specified properties.

THE ANIMAL AND ITS PROPERTIES

It has been shown that an arthropod is an immensely complex organization in which events at the molecular and sub-cellular levels interrelate with those at cellular, tissue, organ, system and whole animal level to give the properties of the viable animal. To specify the detail of the organization absolutely it would be necessary to determine the nature of, to specify the place and the time of occurrence of, and to know the rates of, innumerable events which change and interact minute by minute throughout the life span of the animal. This is impossible and another means of approaching this problem must be found. Consider first the nature of the organism and the terms in which its properties can best be stated.

A fundamental feature is that arthropods (like all other multicellular animals) are constructed so that the genetic information contained within the nucleus is replicated many times, the replicates being separated by only small distances (7–20 μm) throughout the body of the animal. Similarly, the basic features of cellular construction (endoplasmic reticulum, mitochondria, Golgi apparatus and cell membranes), are repeated many times and spread evenly throughout the animal. Much of the complexity of a living organism arises from the multiplicity of events occurring simultaneously in space. In many cells quite different events are occurring in addition to the similar ones. Similarities are due to programme repetition and wide dissemination throughout the organism; differences, to different sets of information in the programme being activated or repressed within the cellular unit.

The mechanisms which cause the activation or repression of the appropriate information in the right cells at the right times are complex and still the subject of much research. Whatever the detail of the mechanisms involved, a useful overall picture may be obtained by likening this arrangement to a choir, where each singer (each cell and nucleus) has a copy of the entire musical work (programme) which he reads and, in accordance with some predetermined plan contained within the programme, puts into action only certain parts of it. Without the whole manuscript he could not put his part in its place and make the correct sequence contribution to the programme. This likeness can be extended: the simultaneous reading of the manuscript by different singers (cells) produces effects not otherwise obtainable from a single singer reading a single programme. Further, while much can be achieved without further elaboration, the presence of a conductor (co-ordinating mechanisms) helps greatly in the proper integration of the singers and may relate their efforts to the mood of the audience (environment).

Every organism then is the product of the decoding of a programme contained within the zygote and expressed in terms of cells, and the environment in which the decoding and transcription occurs. Evolutionary theory proposes ancestral relationships between all known animals, speciation produces barriers in the continuum, and natural selection has led to the survival of some programmes and not of others.

THE RELATIONSHIP OF STRUCTURE, FUNCTION AND DYNAMICS

The terms structure and function are the names given to the products of the first step in the analysis of organization. Structure deals with the manner in which the organism is built; function is what the organism can do by virtue of its structure; and dynamics is what it is actually doing. Of these three properties structure, at an elementary level, can be determined by simple observation and involves little interference with the living organism; function usually requires experimental procedures and interference with the living animal for its determination; dynamics requires very sophisticated procedures for its determination. The more function and dynamics can be related in a deterministic way to structure, the less need there is to interfere with an organism to establish its properties. Structure therefore forms the central theme in discussing these relationships, the more so since structure is the feature which distinguishes arthropod organization from that of all other phyla.

It has been shown that the arthropod body can be analysed into a number of levels, each of which confers certain properties upon the organism. At each level a given structure has a given function and if the structure is known the function can be specified. However more than one kind of structure may have the same function. Experimental analysis may be necessary in the first instance to establish the function of a structure but, once known, the functioning of other animals can be deduced from its occurrence in them. The determination of the structure at one level does not necessarily specify the conditions at the others. If the complete function of an organ is to be deduced from its structure, then the anatomy of the organ, its. tissues, cells and their ultra-structure must be known. Normally however such knowledge is not available and function, if deduced from structure, has to be deduced from fewer facts. This allows the possibility of error, but this can be low and much can be said with truth about the functioning of an animal from few structural observations. Much more error occurs if structure is deduced from function since a given function can be achieved by structures of more than one design.

The momentary dynamic state of the animal is rarely indicated by its structure but its previous dynamic behaviour may leave imprints on its structure: for example, the different anatomies of the forms *gregaria* and *solitaria* of *Locusta migratoria* indicate the previous solitary or gregarious lives led by these animals. Sometimes the accumulation of black pigment will indicate that an animal has had a high metabolic rate for long periods of time.

This refers to living animals which are expected to survive this examination. If, however, the animal is killed and its structure examined in detail, much can be learnt about the past dynamics of the animal, e.g. from the state of the endocrine system, the size of the ovary or the ultra-structure of the cells.

THE EXPRESSION OF ARTHROPOD PROPERTIES

Methods of expressing all the properties of an animal in a single array of facts so that its inherent complexity can be grasped in a meaningful fashion have not been developed. However, omitting to make the attempt would leave the purpose of this book unaccomplished. In what follows a method of expressing the properties of the animal is given. It sets out the important structural, functional and dynamic features of the animal, records the major alternatives that can be found within the Arthropoda, gives a basis for selecting a type of animal organization for a particular purpose, and finally provides a useful record of a set of properties against which the interpretation of any experimental result can be judged. The method clearly springs from matrix analysis in mathematics but the entries in the matrix are biological facts and not real numbers. The manœuvres that can be performed with the mathematical matrix cannot be done with this, the biological matrix. Nevertheless it may be profitable to code the biological facts to give real numbers, and to manœuvre the matrices accordingly. This aspect will not be pursued here but is mentioned to point out to the student that the method, although largely untried, has potentialities well beyond those given, and in the next few pages the basis of very advanced techniques of handling complex situations is outlined.

The array consists of a master-matrix which specifies the main types of information found in an animal and the level at which each influences the organism. Each entry in the master-matrix is itself a matrix which deals with the properties conferred upon the organism by that part of the information specified in the master-matrix and the different major ways in which the properties are expressed. Each entry of the matrix can also be a matrix termed a sub-matrix and can specify further the properties and their manner of expression relevant to the appropriate matrix entry. This can be extended in both directions to give a series of canonical matrices embracing in one direction greater and greater detail about an organism and in the other broader comparisons of different life forms. Here only the master-matrix and its contained matrices will be used to express the array of facts given in the previous chapters about arthropods.

The master-matrix

Each of the four rows of the master-matrix is specified by one of the four sub-programmes into which the information contained within the zygote is sub-divided (Chapter 9). Each sub-programme can be considered as expressing itself in the animal organization at three different levels as follows:

A) Those features in each sub-programme which contrast the arthropod with other life forms at the basic design level.
B) That which influences those features of the animal at the basic design level within the arthropod type.
C) That which influences those features which adjust the animal to some particular mode of life.

Master-Matrix

	A	B	C
C. B. C	C. B. C./A	C. B. C./B	C. B. C./C
G. A. D	G. A. D./A	G. A. D./B	G. A. D./C
F. R. H	F. R. H./A	F. R. H./B	F. R. H./C
A. R. I	A. R. I./A	A. R. I./B	A. R. I./C

These three levels specify the column of the master-matrix, thus giving a matrix of three columns, four rows and twelve entries. Each entry is a code which stands for a group of biological information about the animal which can be allotted to a particular sub-programme at a particular level. The general contents of each entry will now be outlined.

The code is formed from the initials of the appropriate sub-programme followed by a stroke and then the letter **A**, **B**, or **C**, indicating the level at which that content of the sub-programme is influencing the organization of that animal.

C.B.C./A

This contains the properties that specify the chemical and structural basis of the life forms we are studying. The chemical elements and molecules which compose living material and their organization into a cellular structure were determined long before arthropods appeared upon the scene. Nevertheless these features, although common to many animals, do determine much that occurs in arthropods; hence their properties must be noted and included in our studies.

G.A.D./A

This entry deals with the determinants which distinguish the arthropod from all the other patterns which cellular life is known to assume on this planet. The contents of this sub-matrix illustrate the need to use this method of expressing our observations, since it is a set of determinants, not a single one, which gives the life forms we are studying the basic properties that they possess.

F.R.H./A

This entry contains those properties of the frequency, range and homeo-stasis mechanisms which are implicit in the features of the **C.B.C./A** and **G.A.D./A** entries. For example, the segmental nature of the animal and the ladder-like nervous system associated with it lay down the basis for repetitive control mechanisms in each segment and the need to bring them under a central command system; features not basically present in non-segmented animals. The growth and moulting cycle will permit an abruptly increased amount of sensory inflow to the nervous system following an ecdysis, resulting perhaps in a rather more machine-like structure and functioning of the nervous system than if gradual adjustment occurred.

Table 2

C.B.C./A

		1	2	3	4
A	Main chemical constituents (classes of)	a proteins b purines c water	amino acids nucleic acids oxygen	carbohydrates pterins carbon dioxide	lipids inorganic salts
B	Energy source	a radiant b inorganic	chemical organic	anaerobic	aerobic
C	Information coding system	a DNA/RNA	RNA/protein	protein	
D	Unit of organization	a virus	cell	non-cellular	
E	Organelles present	a nucleus b ribosome c Golgi apparatus	nucleolus vacuole plasma membrane	chromosome cilium endoplasmic reticulum	centriole flagellum mitochondrion
F	Basic cell types	a epithelial b sense cell c ciliary d choanocyte	cuticular sperm cell cnidaria storage	muscle ovum collagen solenocyte	neuron mucoid osteoblast musculeo-epithelial

C.B.C./B

	1	2	3	4
A Special biochemical systems	a chitin synthesis b B-vitamin synthesis	arthropodin synthesis	trehalose metabolism	cholesterol synthesis
B Somatic chromosome content x = haploid No.	a 1 N	2 N	4 N	poly N
C Organelles	a tracheole	scolopidia	rhabdome	
D Special cell types	a smooth muscle b fibrillar muscle c rectal papillae d haemocytes e trichogen f amoeboid sperm	striated muscle salivary Malpighian excretory glial neuro-secretory flagellate sperm	tubular muscle silk oenocytes myelinated neurons endocrine	goblet pericardial retinula non-motile sperm

C.B.C./C

	1	2	3	4
A Special chemical products.	a resilin b luciferin c sorbitol	sericin haemoglobin melanin	biopterin formic acid	wax glycerol
B Additional intra-cellular resources	a virus	bacteria	yeast	fungi
C Special cell types	a chromatophores b lipid storage	photophore; glycogen storage	uric acid (storage) protein storage	mycetocytes

G.A.D./A

		1	2	3	4
A	Cellular grade	a unicellular	multicellular	syncytium	
B	Tissue grade	a diploblastic	triploblastic		
C	Integumental surface	a ciliated	epithelial	cuticular	
D	Body symmetry	a spherical	radial	bilateral	pentamerous
E	Body cavity	a absent	blastocoele	coelomic	haemocoele
F	Segmentation	a non-segmented	segmented		
G	Segmental organs	a appendages b coelom c coelomoducts	muscles cardiac chambers	nerve ganglia reproductive	excretory organs nephridia
H	Skeleton	a absent	hydrostatic	internal	exoskeleton
I	Alimentary canal	a mouth	anus		
J	Life pattern	a zygote	egg	larva	adult

G.A.D./B

	1	2	3	4
A Appendages (basic type)	a lobopod b secondarily uniramous	uniramous	trilobite	biramous
B Appendages (specific types)	a pre-antennae b pedipalps c maxilli-pedes d gonopods	antennules mandibles walking limbs cerci	chelicerae maxillae phyllopods	antennae labial swimmerets
C Alimentary canal (organs)	a pharynx b oesophageal valve c rectal sac	gastric mill midgut caeca peritrophic membrane	crop digestive gland	gizzard Malpighian tubules
D Coelomic organs (function)	a salivary	ionic regulation	nitrogen excretion	reproduction
E Respiratory organs	a cutaneous b trachea	gills	lung books	tracheoles
F Brain (basic structure)	a protocerebrum	deutocerebrum	tritocerebrum	other ganglia
G Sense organs	a sensilla trichodea b statocysts c chemoreceptors d compound eye (holocroal)	sensilla campaniformia Johnston's organs contact chemoreceptors compound eye (schizocroal)	sensilla chordotonal tympanic organs ocelli	myochordotonal pressure receptors ommatidia
H Endocrine organs	a neurohaemal b post-commisural c androgenic gland	corpora cardiaca pericardial	corpora allata Y-organ	sinus glands prothoracic glands
I Basic food manipulation system	a microphagous	particulate	liquid	

G.A.D./B (cont'd)

		1	2	3	4
J	Locomotory pattern (basic)	a walking	swimming		
K	Tagmosis (No. in bracket = no. of segments in tagma) (n = unknown or variable seg. no.)	a head (3) / trunk (n)	head (n) / trunk (n) / pygidium (n)	head (6) / trunk (n)	head (6) / thorax (3) / abdomen (11)
		b prosoma (6) / opisthosoma (13) / telson	prosoma (6) / mesosoma (n) / metasoma (n)	head (6) / thorax (n) / abdomen (n) / telson	head (6) / thorax (8) / abdomen (7)
L	Appendage distribution patterns (H = head) (T = trunk) (A = abdomen) (Th = thorax) (P = prosoma) (O = opisthosoma) (M = mesosoma) (m = metasoma) (n = unknown or variable seg. no.)	a pre-antennae H1 / mandibles H2 / slime papillae H3 / trunk limbs Tn	antennae H2 / walking limbs Hn / trunk limbs Tn	antennae H2 / mandibles H4 / maxillae H5 / labial H6 / walking limbs Tr / gonopods T17	antennae H2 / mandibles H4 / maxillae H5 / labial H6 / walking limbs T1–3 / gonopods A8–9 / cerci A11
		b chelicerae P1 / pedipalps P2 / ambulatory P3–6	chelicerae P1 / pedipalps P2 / ambulatory P3–6 / chilaria O1 / genital operculum O2 / plate gills O8–12	antennules H2 / antennae H3 / mandibles H4 / maxillae H5 / maxillae H6 / swimming limbs T1–10 / swimming limbs A3–17	antennules H2 / antennae H3 / mandibles H4 / maxillae H5 / maxillae H6 / walking limbs T1–8 / swimmerets A–
M	Nervous system	a Type I	Type II	Type III	
N	Circulatory system	a Type I	Type II		
O	Alimentary system	a Type I	Type II		
P	Life pattern	a Monophasic	Diphasic	Triphasic	Polyphasic

G.A.D./C

		1	2	3	4
A	Appendage function	*a* sensory *b* locomotory	masticatory grasping	ambulatory adhesion	copulatory
B	Secondary food manipulation system	*a* microphagous	particulate	liquid	blood-sucking
C	Size adult (mm)	*a* 10^3 *b* 10^{-1}	10^2 10^{-2}	10^1	10^0
D	Sex	*a* male	female	worker	hermaphrodite
E	Locomotion	*a* walking *b* burrowing	swimming crawling	running sedentary	flying
F	Life Pattern No. catastrophic ecdyses	*a* none *b* many	one	two	three
G	Distributive phase	*a* young larvae	adult		
H	Ecological niche	*a* prey *b* herbivore	predator inquiline	parasite symbiont	saprophage

In appendage function, the function must be associated with an appendage as listed under **G.A.D./B**, B.

F.R.H./A

	1	2	3	4
A	No. of cell types	*a* 10 to 50	50 to 100	100 to 200	over 200
B	No. of cells per type	*a* tends to minimum	tends to maximum		
C	Gene hormones	*a* present	absent		
D	No. of nerve ganglia in ventral cord (basic)	*a* 3 to 10	10 to 20	20 to 30	over 30
E	No. of instars in life pattern	*a* 3	3 to 7	7 to 14	over 14
F	Duration of life	*a* days	months	year	years
G	Duration of longest stadium	*a* days	weeks	months	years

F.R.H./B

	1	2	3	4
A Tactile sensory field	a cutaneous cerci	antennae median filament	antennule other appendages	pre-antennae
B Chemoreceptive field	a cutaneous palps b other appendages	antennae tarsi	antennules pectines	pre-antennae ovipositor
C Photoreceptive field	a median ocelli b nauplius eye	lateral ocelli	compound eyes	ommatidia
D Sound perception	a trichodea sensilla	tympanic organs		
E Special sensory receptors	a infra-red	vibration	pressure	osmo-receptors
F Number of endocrine organs	a 1 b 5	2 6	3 7	4 8
G Endocrine patterns of instruction	a single hormone signal	continuous single hormone control	relative titer changes of several hormones	
H Patterns of cellular response	a inhibition	stimulation	synchronization	function change

F.R.H./C

			1	2	3	4
A	Metabolic rate. ratio: maximum/minimum	*a*	1	2	3	4
		b	5	6	7	8
B	Median temperature of metabolic rate curve °C	*a*	−5 to +5	+5 to +15	+15 to +25	+25 to +35
		b	+35 to +45	+45 to +55		
C	Water loss over viable range in a dry atms.	*a*	zero	low	moderate	high
D	Photo-sensitive range of spectrum	*a*	ultra-violet	violet	purple	blue
		b	green	yellow	red	infra-red
		c	white			
E	No. of habitats utilized in life pattern	*a*	1	2	3	4
		b	5	6	7	

The following additional entries provide information belonging here, but do not allow the quoting of simple alternatives.

F   The life pattern formula

G   Ecological habitat

H   Geographical range

I   Known stress factors

A.R.I./A

		1	2	3	4
A	No. of nuclei per individual	a 10^6 b 10^{10}	10^7 10^{11}	10^8 10^{12}	10^9 10^{13}
B	No. of segments per individual	a 3–10	10–20	20–30	30

A.R.I./B

		1	2	3	4
A	Sources of alternative information	a alleles	isoenzymes	polyploidy	symbionts
B	Physiological triggers for alternative responses	a hormones	pheromones	neural patterns	other organisms
C	Lability of instar sequence	a normally fixed	altered with difficulty	readily labile	

A.R.I./C

	1	2	3	4
A Diapause stage	a egg	larva	pupa	adult
B Sex determination	a male	female	neuter	
C Caste determination	a gyne	worker	soldier	
D Flexibility of behaviour	a inflexible	modifiable at certain stages only	readily variable	
E Environmental triggers for alternative information	a photoperiod	temperature regime	differential feeding	

A.R.I./A

This entry contains those basic situations, some dependent upon and some independent of arthropod design, which provide the individual with alternative ways of accomplishing the same end (the mechanistic basis of purpose), or provide more than one feature capable of doing the same job, thus guarding against error. The following three examples indicate the type of material which belongs here.

Cellular repetition: the multicellular state provides innumerable copies of the genome. Interactions between the gene products (gene hormones) in different cells show the presence of an error correcting mechanism which might be expected in all animals.

The basic bilateral symmetry of the animal provides a pair of organs, each of which appears capable of doing the whole work of the pair should one fail, or when it is removed experimentally.

Segmentation also increases the number of organs each doing the same job, for example, the coelomic organs. With increasing efficiency of body organization, often only a single pair survives. However, the segmental structure of the animal provides a basis for error protection by repetition of a series of pairs of organs each doing the same job but not working to capacity under normal conditions.

C.B.C./B

The information contained within this matrix relates to the organization of the arthropod cell. It takes those properties given in the **C.B.C./A** matrix and gives them actual expression within the framework of a living multicellular organism, in this case that of an arthropod. Every cell contains not only the information to give it the properties common to all cells, or for cellular features found in some cells but widespread through all Phyla, but also the mechanisms which produce one particular type of supracellular organization and no other, and to which cellular properties are adjusted. Hence, any cell in its entirety is unique to the type of supracellular organization to which it belongs, and it is only some of its properties that are shared with all other animal cells.

This matrix takes note of those cell types which have made possible supracellular organization which is peculiar to arthropods, i.e. tracheolar cells, and of cell types whose absence is peculiar to the Phylum and which exclude its members from lines of organization commonly found elsewhere, i.e. mucus and ciliated cells.

G.A.D./B

This entry deals with the major types of design found in arthropods.

F.R.H./B

This entry deals with those properties of the frequency, range and homeostasis mechanisms which are implicit in the features of the **C.B.C./B** and the **G.A.D./B** entries and additional to those given in the **F.R.H./A** entry. For example, the content of this entry will take account of whether the **G.A.D./B** entry describes an active or a passive type of organism, whether one with a complex or simple endocrine system, or whether the **G.B.C./B** entry specifies one with a wide or limited range of sense cells.

A.R.I./B

In this entry the properties additional to those entered under **A.R.I./A** come from the design types detailed in **C.B.C./B**, **G.A.D./B** and **F.R.H./B** since any set of properties given under these entries may need mechanisms to correct against possible failures, and alternative ways of doing things should environmental circumstances demand it. The most important addition to the sub-programme at this level comes from the growth characteristics of the arthropod. With greater complexity in the life patterns greater differences occur between the structure and function of successive instars. Each instar then has its own specific set of information for its organization. The development of this condition opens the way for control mechanisms which can alter the sequence or repress altogether the appearance of one or more instars according to the requirements of the environment.

C.B.C./C

This entry deals with those cell types and properties which confer upon the individual animal certain abilities such as wax secretion, uric acid storage, chemical detoxication, and chemical defence properties which adjust the animal to some particular way of life, are widespread throughout the arthropod groups but not universally present, and are not involved in the 'B' level of design of the arthropod. Such cells may have appeared independently in the different arthropod evolutionary lines and the identity of, say, wax cells, throughout the group depends only on the assumed similarity of the cellular basis of wax production wherever it occurs.

G.A.D./C

This entry deals with those supracellular characteristics of an arthropod which adapt it for some particular way of life. The sex of the animal is recorded here together with those structures that demand one behaviour pattern rather than another from the individual animal.

F.R.H./C

Much of the dynamics of the animal is included in this entry, such as the metabolic rate of the animal, the frequency of its heart beat, the endocrine control process of body functions, and the extent and nature of its environmental perception. Where possible, actual curves and formulae describing the dynamic processes observed are recorded but often such information is not available. However, some indications can be gained by recording the geographical location and habitat of the animal.

A.R.I./C

This entry deals with the contents of this sub-programme at the level of individual adjustment for some particular mode of life. For example, the ability of the individual arthropod to enter or not enter a diapause state according to environmental requirements, or sex determination, or perhaps more correctly sex realization where it is environmentally controlled, would be properties of the animal to be recorded in this matrix.

The content of the matrices

This general account of the type of information to be included under each entry of the master matrix must now be examined in more detail. This information takes the form of naming the properties of the animal appropriate to each entry, and then listing the different ways in which the property is expressed in the different members of the Arthropoda. The properties to be listed under each entry can be very numerous and different selections could be made according to the purpose of the user. Here the intention is to express at an elementary level the main features possessed by every individual arthropod. Therefore, the properties (a list of which is given in Table II) are derived from the information presented about arthropods in the previous chapters of this book and cover a sufficient range to form a fairly comprehensive description of the group. The manner in which the property is expressed depends upon the nature of the entry being considered. Where possible it will be expressed in anatomical terms for reasons given earlier in the chapter, but in many cases, as for example responses to temperature where a differential or integral equation offers the best expression, or for habitat selection where some ecological classification gives the best description, other forms will be used.

The information arrayed under each entry in the master-matrix is itself best presented in the form of a matrix, where the rows are characterized by the properties and the numbered columns list the means of expressing them. Thus part of entry **G.A.D./B** may read

	1	2	3	4	5	6
Limb type	uniramous	biramous	trilobite	—	—	—

Within any one matrix the numbers of entries in each row will vary with the nature of the property and the number of major ways it is expressed. If, however, the number of columns is kept constant, say equal to the number of ways the most variable property is expressed, it gives a means by which this degree of variability can be recognized, a matter that will appear more important later.

At this elementary level no further elaboration of the matrix entries will be given, but it is possible for each entry itself to have the form of another matrix (sub-matrix). The properties of the Arthropoda are then being expressed through a system of canonical matrices consisting of a master-matrix, expressing the different levels at which the sub-programmes influence the organization of the animal. Each entry of the master-matrix is itself in the form of a matrix and contains the properties present in all animals belonging to the Arthropoda together with all the ways in which they are expressed in the Phylum. Finally, each entry in each matrix could itself be a statement covering an array of facts best expressed as a matrix. Where necessary these can be expressed as sub-matrices and thus a more complex and detailed statement about the Phylum can be made.

This canonical assemblage of matrices shown on page 234 is then a statement about the Phylum Arthropoda. In this form it is not a statement about any lesser group or about an individual animal since only the Phylum can express all the properties so listed. To make from this **Phylum-matrix** one applicable to these more restricted groups, it is necessary to select sets of properties. This can be done in the following manner. A copy of the canonical matrix may be made, leaving the entries of the matrix blank but listing the given properties. Opposite each property may then be entered the selected feature from the Phylum-matrix. Now, some of the matrices will remain unaltered, but, since their contents define properties of the group or animal being considered, they must be included for a complete statement to be made.

The properties and uses of the matrices

This method of setting out the properties of arthropods has several advantages beside the simple documentation of them in an orderly fashion.

1. The same property may be asked for in different matrices. For example, if under **G.A.D.**/**B** the animal is designated as a particulate feeder using mandibular mechanisms and the same question, feeding mechanisms, is asked under the **G.A.D.**/**C** design (adjustments to the way of life), and the selection now specifies liquid feeding, it indicates that considerable evolutionary processes have occurred modifying the primitive mandibulate mechanisms into suctorial ones, as has occurred many times in the Insecta. Where the answers are the same, liquid feeding in both as they would be in Arachnida, then no changes have occurred. A particulate arachnid feeder, if existing, indicates reverse changes. Comparisons therefore between the expression of the properties in the columns of the master-matrix is a good indication of the evolutionary changes that have occurred.

2. Some of the matrices will remain unchanged, although others change. Thus, where adjustments to life within a group are being considered, the contents of columns **A** and **B** in the master-matrix may well remain unchanged; variations in **C** can express all the adjustments necessary. Similarly, **A** and **C** could well remain unchanged if **B** is selected for large groups to give a sub-phylum or Class matrix where the group has penetrated into such a range of habitats that it is no more restricted in this respect than is the Phylum.

3. Once the matrix has been constructed a plan is produced indicating the many properties of the group so described. For individual animal matrices, this plan can be used in helping to interpret experimental results where factors outside the experiment have to be taken into account to arrive at an understanding of facts so gained.

4. The matrix may aid in the selection of an experimental animal for a particular purpose, since very often the interpretation of an experiment is aided or made more difficult by properties of the animal not directly concerned in the experimental conditions. For example, the length of life history, the size of the animal, or the possibility that the property studied is secondarily rather than primarily present. These specifications may need only the scrutiny of one matrix. Constancy in the others would indicate the wide range of animals that may be found that fill the requirements of the experimenter.

To find an arthropod which has the properties specified in any coherent matrix is a matter of relating the appropriate features of the matrix with the keys and description given in the various taxonomic works on the Arthropoda. Some start has been given in this book, so that an appropriate selection of factors in **G.A.D.**/**B** and their relationship to the Orders of arthropods as given in Chapters 6 and 9, makes identification to this level fairly straightforward. Identification at the other levels can only come with experience and increasing knowledge of the many types of arthropods documented in the literature. However, the properties listed in the matrix are not recorded in full for more than a handful of species, and much work needs to be done before a type specified can be traced through the existing descriptions and known properties.

There are however other ways. The listing of coherent sets of properties specifies the animal fairly closely and it should be possible to form a fair idea where such an animal may be found, and it may well be quicker to go and look for it and to run selection tests upon likely candidates than to trace possibilities through the many languages and wide field of literature in which they may be recorded.

Finally, it may be possible to produce by crossing, mutation or experimental embryology the type that is required. Indeed the study of animal organization is very much in need of methods by which the properties and parts of organisms can be altered to give organisms with anatomies different (and not by mere subtraction) from those normally present.

THE NATURE OF ARTHROPOD ORGANIZATION

The name Arthropoda is given to one of the supracellular designs that have evolved in the world's fauna; one which, if the existence of large numbers of individuals, large numbers of species, and the ability to adapt to nearly all the earth's habitats are taken as criteria, is one of the most successful ones, if not the most successful. No other design of comparable complexity exists which occupies the physical size range lying between the small interstitial sand copepods, fractions of a millimeter in size, to the 3–4 metre *Macrocheira kämpferi* crabs from the Japanese seas. The actual size range, not the relative one, is the important thing to note here. To what feature does this design owe its success? What follows is necessarily rather speculative and the view given of the basic nature of arthropod organization is opinion rather than fact.

One of the noticeable attributes of arthropods is the small numbers of cells which make up their tissues and organs. For example, the muscles may be composed of only a few cells, sometimes only a single fibre; the epithelia are usually only one cell thick; chemoreceptors may be formed from only four cells each performing a special function, while integrative centres within the central nervous system may be single neurons rather than groups of neurons. This is not just a matter of absolute size since a comparison between say a small rodent and an arthropod of equal size will show just these differences. In the rodent the muscles are made of thousands of cells; all the epithelia are many cells thick; the chemoreceptors are composed of tens of cells and the integrative centres of the nervous system of thousands of neurons. This is a difficult comparison to express in figures or to prove by detailed comparison between the different Phyla, but the arthropod design does appear to be one which tends to use the minimum possible number of cells in its organization. It is also one of considerable anatomical complexity, so that there is a high degree of differentiation amongst the cells found in these animals. This gives a design built upon a fair number of cell types, each of which is present in the minimal possible quantity.

It has been shown that the performance of any function by a cell takes time and that very often cells performing the same function may not be closely synchronized in their actions. If a large number of cells of the same type are present, all stages in the functions are likely to be present amongst such a population at any one time. If however, the number of cells of a given type is less than the number of possible states through which the function can progress, some states will be absent from the population at any one time. This can be extended to the number of possible functions that the cell can perform; again, if this exceeds the number of cells present, not all functions of these cells can be expressed in that population at any one time. At some stage then this tendency to minimal cell number will reach a level where only limited sets of properties can be realized in that cell population at any one time, a condition which can lead to a rather machine-like animal capable at any one time of responding to the environment in a limited and stereotyped manner. It seems probable that many of the arthropods, particularly the smaller ones, are in this condition. They are quite complex animals with many cell types each represented by minimal quantities of cells, giving an animal organization of limited and stereotyped responses.

Now within the arthropods there seem to be two ways by which the animal can escape from these limitations.

The individual animal, while it may not be able to exhibit a large number of functions in a small number of cells at any one time, may well be able to do so at different times; indeed, the many functions displayed in time by the epithelial cells illustrate this possibility. There are two requirements for this possibility. The information to change the cell function from one thing to another must be held in the cell, usually in the genotype, and there must be some mechanism to call it forth when required. This implies a rather strong link between the stored information and the events occurring within the cell, and the arthropod, especially the insect, appears to have its organization and its life pattern more fully programmed from its genome than appears the case in other animal designs.

The second method of escape is to produce a population of *different* individuals, each however of limited adaptability. In such a population some at least of the individuals will possess the stereotyped responses which will fit the conditions imposed upon them and the population will survive. All animal groups show this phenomenon but in arthropods it appears to be particularly well developed. Thus arthropod species tend to be made up of numerous populations each of which has individuals showing genetic adaptations for the particular requirements of their habitats, together with a proportion which are not so well adapted. Also, if the generation interval is short enough, cycles of adaptation to the seasonal changes throughout the years may be shown in the genotype of the population.

The large extent to which the organization and the life pattern of the animal are programmed in the genetic system, together with the need for many variations of this programme to be presented by different individuals of the population, gives the basis for genetic incompatibility between some members of the population and between populations. Thus sterility barriers may be quickly formed and perhaps just as quickly destroyed. Often, perhaps more often than in other groups, such barriers will have a long duration allowing many differences to accumulate between such populations. This may well be reflected by the large number of species which exist in arthropods, more than in all the other groups put together.

This basic nature of the arthropod of many cell types each present in minimal quantities giving stereotyped individuals whose responses are extensively programmed in the genome, is expressed in a supracellular anatomy whose properties have been described and summed up in the Phylum-matrix already given. At this level also the basic nature of arthropod design is one of increased size and complexity with minimal variation, as is shown by the increase in number of similar segments with their similar organs, organization and appendages. Minimal differences occur to adapt the anterior and posterior segments for their special functions. The basic evolution of the body organization of the Phylum has largely been in ways by which these limitations can be overcome. Sequential changes in function of the cells may not extend the ability of the animal at any one moment, but given time, the animal can adapt to a wide range of circumstances, perhaps wider than is shown by any other group of animals. By increasing the numbers of kinds of individuals that can be thrown into a population, stereotyped animals of suitable types will occur.

The success of arthropods, then, springs from the basic features of their organization; of a body having many diverse cell types each present in minimal numbers, linked to detailed programming within the zygote of the individual's organization and life pattern and possible responses. The restrictive features of this type of organization are partially or completely overcome by sequential changes of cell function and/or body organization in time, and by the production within a population of large numbers of individuals each with a different programme.

This type of organization is also one which is very advantageous to a small animal, since complexity and small physical size are naturally linked with diverse cell types present in minimal numbers, and the relatively stable conditions within a micro-habitat are less disadvantageous to a highly programmed stereotyped animal than would be the more variable ones found in macro-habitats. It is just over this size range that the arthropods are the supreme animals of the world's fauna.

Bibliography

BALDWIN, E. (1940). *An Introduction to Comparative Biochemistry.* Cambridge University Press.

BARGMANN, W. (1966). Neurosecretion, *Int. Rev. Cytol.*, **19**, 183–201.

BARTH, R. H. JR. (1967). The Comparative Physiology of Reproductive Processes in Cockroaches. Part I. Mating behaviour and its endocrine control, *Adv. reprod. Physiol.*, **3**, 167–207.

BARTON-BROWNE, L. B. (1964). Water regulation in insects, *Ann. Rev. Entomol.*, **9**, 63–82.

BERNHARD, C. G. (1966). The functional organization of the compound eye, *Proc. int. Symp.* Stockholm, Pergamon Press, Oxford.

BULLOCK, T. H. and HORRIDGE, G. A. (1965). *The Structure and Function of the Nervous Systems of Invertebrates, I and II,* W. H. Freeman, Reading, Berks.

BUTLER, C. G. (1954). *The World of the Honey Bee,* Collins, London.

BUTLER, C. G. (1967). Insect pheromones, *Biol. Rev.*, **42**, 42–87.

BUTT, F. H. (1960). Head development in arthropods, *Biol. Rev.*, **35**, 43–91.

CANNON, H. G. (1933). On the feeding mechanism of the Branchiopoda, *Phil. Trans. R. Soc.*, B, **222**, 267–352.

CARLISLE, D. B. and KNOWLES, Sir FRANCIS (1959). *Endocrine Control in Crustaceans,* Cambridge University Press.

CARTHY, J. D. (1965). *The Behaviour of Arthropods,* Oliver and Boyd, Edinburgh.

CHRISTOPHERS, S. R. (1960). Aedes Egypti (*L.*), *the Yellow Fever Mosquito. Its Life History, Bionomics and Structure,* Cambridge University Press.

CLARKE, K. U. (1967). Insects and temperature. In *Thermobiology,* ed. A. H. Rose, pp. 293–352. Academic Press, London and New York.

CLEVER, U. (1968). Regulation of chromosome function, *Ann. Rev. Genet.*, **2**, 11–30.

COTT, H. B. (1940). *Adaptive Coloration in Animals*, Methuen, London.

COUNCE, S. J. (1961). The analysis of insect embryogenesis, *Ann. Rev. Entomol.*, **6**, 295–312.

DA CUNHA, A. B. (1960). Chromosomal variation and adaptation in insects, *Ann. Rev. Entomol.*, **5**, 85–110.

DANILEVSKĮI, A. S. (1965). *Photoperiodism and Seasonal Development of Insects*, Oliver and Boyd, Edinburgh.

DAVEY, K. G. (1965). *Reproduction in the Insects*, Oliver and Boyd, Edinburgh.

DEMEREC, M. (1950). *Biology of* Drosophila, John Wiley, New York; Chapman and Hall, London.

DETHIER, V. G. (1963). *The Physiology of Insect Senses*, Methuen, London; John Wiley, New York.

DOWNES, J. A. (1965). Adaptations of insects in the arctic, *Ann. Rev. Entomol.*, **10**, 257–274.

ECCLES, J. C. (1964). *The Physiology of Synapses*, Academic Press, New York and London.

EDWARDS, J. S. (1969). Postembryonic development and regeneration of the insect nervous system, *Adv. Insect Physiol.*, **6**, 97–137.

EWING, A. W. and MANNING, A. (1967). The evolution and genetics of insect behaviour, *Ann. Rev. Entomol.*, **12**, 471–494.

GABE, M. (1966). *Neurosecretion*, International Series of Monographs in Pure and Applied Biology. Kerkut, G. A. (ed.), Pergamon Press, Oxford.

GILMOUR, D. (1961). *The Biochemistry of Insects*, Academic Press, New York and London.

GILMOUR, D. (1965). *The Metabolism of Insects*, Oliver and Boyd, Edinburgh.

GRASSÉ, P. P. (1949–1951). *Traité de Zoologie, Anatomie, Systématique, Biologie, Insectes.*, Mason, Paris.

GRAY, Sir JAMES (1968). *Animal Locomotion*, Weidenfeld and Nicolson, London.

GURNEY, R. (1942). *Larvae of Decapod Crustacea*, The Ray Society, London.

HAAS, H. J. (1968). On the epigenetic mechanisms of patterns in the insect integument (A reappraisal of older concepts), *Int. Rev. Gen. exp. Zool.*, **3**, 2–51.

HARRIS, W. V. (1961). *Termites, their Recognition and Control*, Longmans, London.

HASKELL, P. T. (1961). *Insect Sounds*, H. F. and G. Witherby, London.

HERMAN, W. S. (1967). The ecdysial glands of arthropods, *Int. Rev. Cytol.*, **22**, 269–349.

HIGHNAM, K. C. (1962). Neurosecretory control of ovarian development in the desert locust. In *Neurosecretion* (H. Heller and R. B. Clark, eds.), pp. 379–389. *Memoirs Soc. Endocrinol.*, **12**.

HIGHNAM, K. C. and HILL, L. (1969). *The Comparative Endocrinology of the Invertebrates*, Edward Arnold, London.

HINTON, H. E. (1958). The phylogeny of the Panorpoid Orders, *Ann. Rev. Entomol.*, **3**, 181–206.

HODGSON, E. S. (1958). Chemoreception in arthropods, *Ann. Rev. Entomol.*, **3**, 19–36.

HOHN, F. E. (1960). *Applied Boolean Algebra. An Elementary Introduction*, Macmillan, New York.

HUGGINS, A. K. and MUNDAY, K. A. (1968). Crustacean metabolism, *Adv. comp. Physiol. Biochem.*, **3**, 271–378.

HUXLEY, J. S. (1932). *Problems of Relative Growth*. Methuen, London.

IMMS, A. A. (1957). *A General Textbook of Entomology*, rev. ed. (9th), Richards and Davis, Methuen, London.

JEANNEL, R. (1960). *Introduction to Entomology*, trans. H. Oldroyd, Hutchinson Scientific and Technical, London.

JOHANNSEN, O. A. and BUTT, F. H. (1941). *Embryology of Insects and Myriapods*, McGraw-Hill, New York.

KENNEDY, D. (1966). The comparative physiology of invertebrate central neurons, *Adv. comp. Physiol. Biochem.*, **2**, 117–184.

KROEGER, H. and LEZZI, M. (1966). Regulation of gene action in insect development, *Ann. Rev. Entomol.*, **11**, 1–22.

LANKESTER, E. R. (1885). *Limulus*, an arachnid, *Q.Jl. microsc. Sci.*, **21**, 504–548 and 609–649.

LEDLEY, R. S. (1965). *The Use of Computers in Biology and Medicine*, McGraw-Hill, New York.

LEES, A. D. (1955). *The Physiology of Diapause in Arthropods*, Cambridge University Press.

LOCKE, M. (1967). The development of patterns in the integument of insects, *Adv. Morphogen.*, **6**, 33–88.

LOCKWOOD, A. P. M. (1968). *Aspects of the Physiology of Crustacea*, Oliver and Boyd, Edinburgh.

MANTON, S. M. (1937). The feeding, digestion, excretion and food storage of *Peripatus*, *Phil. Trans. R. Soc.*, B, **227**, 411–464.

MANTON, S. M. (1949). Studies in Onychophora VII. The early embryonic stages of *Peripatus* and some general considerations concerning morphology and phylogeny of the Arthropoda, *Phil. Trans. R. Soc.*, B, **233**, 483–580.

MANTON, S. M. (1960). Concerning head development in arthropods, *Biol. Rev.*, **35**, 265–282.

MANTON, S. M. (1964). Mandibular mechanisms and the evolution of arthropods, *Phil. Trans. R. Soc.* B, **247**, 1–184.

MITTELSTAEDT, H. (1962). Control systems of orientation in insects, *Ann. Rev. Entomol.*, **7**, 177–198.

NEVILLE, A. C. (1967). Chitin orientation in cuticle and its control, *Adv. Insect Physiol.*, **4**, 213–286.

du NOUY, L. (1936). *Biological Time*, Methuen, London.

du PORTE, E. M. (1957). The comparative morphology of the insect head, *Ann. Rev. Entomol.*, **2**, 55–70.

PRINGLE, J. W. S. (1957). *Insect Flight*, Cambridge University Press.

PUNNETT, R. C. (1915). *Mimicry in Butterflies*, Cambridge University Press.

RICHARDS, A. G. (1951). *The Integument of Arthropods*, University of Minnesota Press, Minneapolis.

RICHARDS, O. W. (1953). *The Social Insects*, MacDonald, London.

ROCKSTEIN, M. (1964–1965). *The Physiology of Insecta*, Vols. I, II and III, Academic Press, New York and London.

ROEDER, K. D. (1953). *Insect Physiology*, John Wiley, New York; Chapman and Hall, London.

ROEDER, K. D. (1963). *Nerve Cells and Insect Behaviour*, Harvard University Press.

ROSEN, R. (1967). *Optimality Principles in Biology*, Butterworth, London.

ROSEN, R. (1968). Recent developments in the theory of control and regulation of cellular processes, *Int. Rev. Cytol.*, **23**, 25–88.

ROTH, L. M. and EISNER, T. (1962). Chemical defences of arthropods, *Ann. Rev. Entomol.*, **7**, 107–136.

ROTHSCHILD, Lord (1965). *A Classification of Living Animals*, Longmans, London.

SALT, R. W. (1961). Principle of insect cold-hardiness, *Ann. Rev. Entomol.*, **6**, 55–74.

SCHARRER, E. and SCHARRER, B. (1963). *Neuroendocrinology*, Columbia University Press, New York.

SHAROV, A. G. (1966). *Basic Arthropodan Stock with Special Reference to Insects*, Pergamon Press, Oxford.

SHAW, C. R. (1969). Isoenzymes: classification, frequency and significance, *Int. Rev. Cytol.*, **25**, 297–332.

SLIFER, E. H. (1970). The structure of arthropod chemoreceptors, *Ann. Rev. Entomol.*, **15**, 121–142.

SMITH, D. S. (1968). *Insect Cells. Their Structure and Function*, Oliver and Boyd, Edinburgh.

SNODGRASS, R. E. (1935). *Principles of Insect Morphology*, McGraw-Hill, New York.

SNODGRASS, R. E. (1952). *A Textbook of Arthropod Anatomy*, Comstock Publishing Association, Ithaca, New York.

SNODGRASS, R. E. (1959). *Studies in Invertebrate Morphology*, Smithsonian Institution, Washington, D.C.

THOMPSON, D. W. (1952). *On Growth and Form*, Vols. I and II, Cambridge University Press.

TIEGS, V. W. and MANTON, S. M. (1958). The Evolution of arthropods, *Biol. Rev.*, **33**, 255–337.

TREHERNE, J. E. (1966). *The Neurochemistry of Arthropods*, Cambridge University Press.

VANDEL, A. (1965). *Biospeliology. The Biology of Cavernicolous Animals*, Pergamon Press, Oxford.

WATERMAN, T. H. (1960) *The Physiology of Crustacea*, Vol. I *Metabolism and Growth*, Academic Press, New York and London.

WATERMAN, T. H. (1961). *The Physiology of Crustacea*, Vol. II *Sense Organs and Behaviour*, Academic Press, New York and London.

WELLS, M. J. (1965). Learning by marine invertebrates, *Adv. marine Biol.*, **3**, 1–62.

WHITTINGTON, H. B. and ROLFE, W. D. (1963). *Phylogeny and Evolution of Crustacea*, Museum of Comparative Zoology, Spec. Publ. i–xi, 1–192, Cambridge, Mass.

WIGGLESWORTH, V. B. (1964). *The Life of Insects*, Weidenfeld and Nicolson, London.

WIGGLESWORTH, V. B. (1959a). Insect blood cells, *Ann. Rev. Entomol.*, **4**, 1–16.

WIGGLESWORTH, V. B. (1959b). *The Control of Growth and Form. A study of the epidermal cell in an insect*, Cornell University Press, Ithaca, New York; Oxford University Press, London.

WIGGLESWORTH, V. B. (1965). *The Principles of Insect Physiology*, 6th ed., Methuen, London.

de WILDE, J. (1962). Photoperiodism in insects and mites, *Ann. Rev. Entomol.*, **7**, 1–26.

WILSON, D. M. (1968). The nervous control of insect flight and related behaviour, *Adv. Insect Physiol.*, **5**, 289–338.

WOOLEY, T. A. (1961). A review of the phylogeny of mites, *Ann. Rev. Entomol.*, **6**, 263–284.

Systematic Index

Subject Index

Figures in bold type denote illustrations

protocerebrum, 79, 154, 165
pupa, 136–7, 143, 146, 157–8

races, 210–211, 212, 213, 214
relationships between systems, 109–12,
110
reproductive systems, 104–7, 105, 106
resilin, 196
respiration,
nervous control of in insects, 162–4,
164
respiratory organs, 52, 50–3
air sacs, 53, 52, 101
gills, 51, 52, 98, 101, 101, 111
lung books, 51, 52, 98, 101, 101, 111
plastron, 53
spiracles, 51, 52, 53, 101, 102
trachea, 52, 53, 67, 101, 101
tracheal gills, 53
tracheoles, 51, 52, 101
respiratory system, 98–102
rhodopsin, 10

salivary glands, 54, 54, 73
sclerites, 49, 50, 90, 116, 118
sclerotic plates, 49, 50, 116
secondary segmentation, 91, 92
segmentalization, 88–90
segmentation, 88, 91
embryonic development of, 126–7
segments, evolutionary history of, 172–
174
sense organs, 56–66, 205–7
chemoreceptors, 61–2, 61
chordotonal sensilla, 57, 58, 60
eyes, 62–66, 63, 205–7, 218
compound eye, 64–6, 65
holocroal eye, 64, 66, 65
ocelli, 62, 205, 206, 207
ommatidium, 62
schizocroal eye, 64, 66, 65
Johnston's organ, 59, 57
mechanoreceptors, 59–61
myochordotonal organs, 58
sensilla campaniforma, 57, 57
sensilla trichodea, 57, 60
sensillum basonicum, 61
statocyst, 57, 59
statolith, 59
typanic membranes, 60
tympanic organs of insects, 60

sequential processing, 67
service cells, 22
sinus gland, 153, 158–60
skeletal plates, 50, 91
skeletal systems, 90–5, 91
Snodgrass, (scheme of classification),
189, 258
spermatophore, 105–6
spiracles, 51, 52, 53, 101, 102
stadium, 131
stomatogastric ganglia, 85
structure, function and dynamics, relationship of, 230–1
sub-chelae, 46
Sub-classes, 188, 190–2
Super-orders, 188
swimming in Crustacea, 48
synapsis, 28, 29
synaptic vesicles, 19, 19

temperature, effect on growth and
development, 143–8
tissues, 22–32
main types of, 22, 23
epithelial, 22–7, 23, 24
epidermis, 22, 25–6, 26, 27
cuticle, 25–6, 26, 27, 28, 35, 90,
207–8
three-dimensional types, 22, 27
nerve tissues, 27–31
open tissues, 31–2
blood tissues, 31–2
muscle tissues, 31–2
tormogen, 56
trachea, 52, 53, 67, 101, 101
tracheal gills, 53
tracheoles, 51, 52, 101
trichogen, 56
triphasic life histories, 142–3, 142
tubular organs, 67–79
distributive organs, 67, 77–9
blood vessels, 67, 79
heart, 67, 77, 77, 78, 102, 159
trachea, 52, 53, 67
organs of sequential processing, 67–
77
coelomic organs, 67, 73–7, 74,
104
gonads, 75–7, 76
gut, 67–73, 68, 100, 104, 209–10
tympanic membranes, 60
tympanic organs, 60